Power Systems

Electrical power has been the technological foundation of industrial societies for many years. Although the systems designed to provide and apply electrical energy have reached a high degree of maturity, unforeseen problems are constantly encountered, necessitating the design of more efficient and reliable systems based on novel technologies. The book series Power Systems is aimed at providing detailed, accurate and sound technical information about these new developments in electrical power engineering. It includes topics on power generation, storage and transmission as well as electrical machines. The monographs and advanced textbooks in this series address researchers, lecturers, industrial engineers and senior students in electrical engineering.

**Power Systems is indexed in Scopus**

More information about this series at http://www.springer.com/series/4622

Ibrahim Moukhtar · Adel Z. El Dein ·
Adel A. Elbaset · Yasunori Mitani

# Solar Energy

## Technologies, Design, Modeling, and Economics

Springer

Ibrahim Moukhtar
Faculty of Energy Engineering
Electrical Engineering Department
Aswan University, Aswan, Egypt

Adel A. Elbaset
Faculty of Engineering
Electrical Engineering Department
Minia University, El-Minia, Egypt

El-Arish High Institute for Engineering
and Technology
El-Arish, North Sinai, Egypt

Adel Z. El Dein
Faculty of Energy Engineering
Electrical Engineering Department
Aswan University, Aswan, Egypt

Yasunori Mitani
Electrical and Electronic Engineering
Kyushu Art Institute of Technology
Tobata-ku, Fukuoka, Japan

ISSN 1612-1287 ISSN 1860-4676 (electronic)
Power Systems
ISBN 978-3-030-61306-8 ISBN 978-3-030-61307-5 (eBook)
https://doi.org/10.1007/978-3-030-61307-5



This Springer imprint is published by the registered company Springer Nature Switzerland AG
The registered company address is: Gewerbestrasse 11, 6330 Cham, Switzerland

*Dedicated to our families and students*

# About This Book

Solar energy systems such as Photovoltaic (PV) systems and Concentrated Solar Power (CSP) technologies are one of the most important Renewable Energy Sources (RES) that convert solar energy into electricity in an environment-friendly manner and a free energy source. Recently, among all of the CSP systems types, the Central Tower Receiver (CTR) technology draws extensive attention as a promising candidate for large solar thermal plants because of its high efficiency resulting from high operating temperature. Indeed, the merits and cost reduction of solar energy systems, especially PV systems, led to a rapid increase in the integration of PV systems over the past decades. However, the high penetration level of the solar PV system has significant effects on the system's power quality due to the mismatch between demand patterns and solar resources. These negative impacts limit the high penetration levels of PV systems. Therefore, PV itself may represent a new challenge to the electrical system rather than being a part of the solution. Furthermore, the design and modelling of solar energy systems, particularly Central Tower Receiver Power Plants (CTRPP), have many difficulties if all technical parameters are taken into account.

In this book, a proposed design was used to satisfy the power demand, get the minimum Levelized Cost of Electricity (LCOE), and get an optimal design for PV and CTR systems. In addition, this work addresses mathematical modelling for CTRPP with Thermal Energy Storage (TES) from a reasonably simplified model perspective. This simplified model is appropriate for studying power system's reliability. Furthermore, a facile controllable scheme based on an Artificial Neural Network (ANN) technique has been adopted for estimating the discharge rate of Heat Transfer Fluid (HTF) from the Cold Storage Tank (CST) and consequently control the receiver outlet temperature. Three different ANN models, such as Radial Basis Function (RBF), Generalized Regression Neural Network (GRNN), and Multi-layer Perceptron (MLP), were applied to assess the performance of the proposed model of CTR-ANN plant. As well, three learning algorithms, Levenberg-Marquardt (LM), Quasi-Newton (BFG), and Scaled Conjugate Gradient (SCG), were selected for training the Multi-layer Perceptron Model. The performance of the adopted CTR-ANN model was presented and compared to the System Advisor Model (SAM) results.

Finally, in this book, the impact of CTR/TES technology on the penetration of PV systems without electrical batteries has been analysed. Based on statistical error analysis, it is found that the MLP model with LM learning algorithms is the optimal model compared to other models. Therefore, the MLP gives a precise estimation for the HTF discharge rate. Hence, the receiver outlet temperature remains constant at the desired value regardless of the variations in direct solar radiation and receiver inlet temperature. The simulation results exhibit that the adopted the CTR-ANN model, and SAM results are in good agreement with each other. The reasonable simplicity and minimum required input data of CTR-ANN model make it an adequate tool to predict and analyse the performance of CTR technology in a simple and fixable manner. The results proved that the reduction in LCOE through 2030 for the CTR plant is significantly greater than PV due to technology improvements and other factors such as reduction of TES cost. Therefore, the CTR systems might become the technology of choice in the near future. Additionally, the results show that the performance of CSP has a great positive impact on the system reliability and the penetration level of PV systems. CSP supports the overall system flexibility due to its ability to store and dispatch the generated power. Additionally, CSP reduces the minimum generation constraint of the conventional generators that allow more penetration of the PV system.

# Contents

# Abbreviations and Symbols

## Abbreviations

| | |
|---|---|
| ANN | Artificial neural network |
| BFG | Quasi-Newton |
| BoS | Balance of system |
| BPC | Balance of plant cost per kW |
| BSC | Balance of system cost, $\$/W_{dc}$ |
| CC | Contingency cost |
| CF | Capacity factor |
| CoI | Cost of inverter |
| COV | Coefficient of variation |
| CRF | Capital recovery factor |
| CSP | Concentrated solar power |
| CST | Cold storage tank |
| CTR | Central tower receiver |
| CTRPP | Central tower receiver power plant |
| DCC | Direct capital cost |
| DI | Daily integration |
| DNI | Direct normal irradiation |
| EDO | Engineering and developer overhead |
| EPC&OC | Engineer-procure-construct and owner costs |
| FCPV | Fixed cost by capacity of PV system, $/kW-yr |
| FTC | Fixed tower cost |
| GI | Grid interconnection |
| GRNN | Generalized regression neural network |
| HFC | Heliostat field cost |
| HST | Hot storage tank |
| HTF | Heat transfer fluid |
| ICC | Indirect capital cost |
| IEA | International Energy Agency |
| IL | Installation labor |

| | |
|---|---|
| ILC | Installation labor cost, $\$/W_{dc}$ |
| LAC | Land area cos of PV system, \$/acre |
| LCOE | Levelized cost of electricity |
| LCoPV | Land cost of PV system |
| LF | Linear Fresnel |
| LM | Levenberg-Marquardt |
| LPC | Land preparation cost of PV system, $\$/W_{dc}$ |
| MAE | Mean absolute error |
| MC | Monte Carlo |
| MENA | Middle East and North Africa |
| MLP | Multi-layer perceptron |
| MSE | Mean square error |
| NTU | Number of transfer units |
| $NTU_{ref}$ | Number of transfer units at the reference full-load condition |
| OMCPV | Operation and maintenance costs of PV system |
| ORC | Organic Rankine cycle |
| PB | Power block |
| PBC | Power block cost per kW |
| PCC | Power cycle cost |
| PD | Parabolic dish |
| PDC | Percentage of Direct Cost of PV system |
| PES | Permitting—Environmental Studies |
| PT | Parabolic trough |
| PV | Photovoltaic |
| PVDC | Direct cos of PV system |
| RBF | Radial basis function |
| RCSE | Receiver cost scaling exponent |
| RES | Renewable energy sources |
| RMSE | Root mean square error |
| RRC | Receiver reference cost |
| SAH | Solar air heater |
| SAM | System Advisor Model |
| SCG | Scaled conjugate gradient |
| SF | Solar field |
| SG | Steam generator |
| SIC | Site Improvements cost |
| SM | Solar multiple |
| SPC | Site preparation cost per unit area |
| ST | Sales tax |
| STR | Sales tax rate of PV system |
| TCSE | Tower cost scaling exponent |
| TES | Thermal energy storage |
| TESC | Thermal energy storage cost |
| TIC | Total inverter cost |
| TLAC | Total land area cost |

| | |
|---|---|
| TMC | Total modules cost |
| TRC | Total receiver cost |
| TTC | Total tower cost |
| VCPV | Variable cost by generation of PV system, $/MWh |

## Symbols

| | |
|---|---|
| $\alpha_s$ | Solar altitude angle (degree) |
| $h_s$ | Solar hour angle (degree) |
| $\delta_s$ | Solar declination angle (degree) |
| $\gamma_s$ | Solar azimuth angle (degree) |
| $\theta z$ | Solar zenith angle (degree) |
| $\beta$ | Surface slope angle (degree) |
| $\gamma$ | Surface azimuth angle (degree) |
| $\theta_{inc}$ | Incidence beam angle (degree) |
| $L_{lat}$ | Local latitude (degree) |
| $L_{long}$ | Local longitude (degree) |
| $L_s$ | Standard time meridian (degree) |
| $t_s$ | Solar time (hour) |
| $t$ | Local time (hour) |
| $E$ | Equation of time (minute) |
| $N$ | Day number during the year |
| $h_{ss}$ | Hour angles of sunset (degree) |
| $A(h_{ss})$ | Factor depends on hour angles of sunset |
| $B(h_{ss})$ | Factor depends on hour angles of sunset |
| $a_2/a_1$ | Atmospheric extinction effect |
| $S_d$ | Day length (hour) |
| $K_t$ | Daily average clearness index |
| $H_o$ | Daily-average extraterrestrial irradiation on a horizontal surface (J/m$^2$ day) |
| $I_o$ | Solar constant, 1366.1 W/m$^2$ |
| $R_{cf}$ | Monthly average sun–earth correction factor |
| $h_o$ | Daily average solar elevation outside of atmosphere (degree) |
| $q$ | Factor depends on location site and declination angle |
| $\hat{H}_h$ | Long-term average daily total irradiation on a horizontal surface (J/m$^2$ day) |
| $\hat{H}_d$ | Long-term average daily diffuse irradiation on a horizontal surface (J/m$^2$ day) |
| $r_t$ | Ratio of hourly total to the long-term average daily total irradiation on a horizontal surface |
| $r_d$ | Ratio of the hourly diffuse to the long-term average daily diffuse irradiation on a horizontal surface |
| $\rho_g$ | Ground reflectance factor |

| | |
|---|---|
| $I_{sol}$ | Solar radiation (W/m$^2$) |
| $r/h_t$ | Non-dimensional ratio between the radial distance from the tower and tower height |
| $h_t$ | Tower height (m) |
| $\theta_{ref}$ | Reflected beam angle (degree) |
| $R_{HR}$ | Heliostat to receiver unit vector |
| $S_{HS}$ | Heliostat to sun unit vector |
| $v_o$ | Location of an aim point, $A_T$, above the origin point (m) |
| $v_1$ | Height of a heliostat mirror, H, above the origin point (m) |
| $x_1$ and $y_1$ | Location of a the heliostat mirrors, H, in east and north direction, respectively (m) |
| $\eta_{cos}$ | Factor of cosine effect |
| $\eta_{cos,fr}$ | Fractional annual cosine factor |
| $cos\theta_{h,p}$ | Cosine factor at each point of the heliostats field for each hour in the year |
| $P_m$ | Annual reflected solar power per mirror area (MW) |
| $P_{LA}$ | Annual reflected power per unit land area (MW) |
| $I_{sol,h}$ | Solar radiation for each hour in the year (W/m$^2$) |
| $P_{e,d}$ | Design capacity (MW) |
| $P_{th,d}$ | Design thermal power of the HTF input to the PB (MW) |
| $P_{rec,d}$ | Required receiver solar thermal power that collected by the heliostat field (MW) |
| $P_{rec,h}$ | Receiver solar power that reflected from the heliostats field at any hour (MW) |
| $dx/h_t$ | Dimensionless 'x' coordinate of a point on the field |
| $dy/h_t$ | Dimensionless 'y' coordinate of a point on the field |
| $\eta_{pb}$ | PB efficiency |
| $N_h$ | Heliostats number |
| $\eta_{refl}$ | Efficiency of mirror reflection |
| $\eta_{att}$ | Attenuation factor |
| $\eta_{att,p}$ | Attenuation factor at each field point |
| $h_{t,ini}$ | Initial tower height (m) |
| $S_{n,p}$ | Slant height of the point, *p*, from the top of the tower (km) |
| $h_{t,f}$ | Final tower height (m) at SM = one |
| $dh_t$ | Tower height increment (m) |
| $h_{t,SM}$ | Tower height (m) at any SM value |
| $P_{HE,h}$ | Actual input solar thermal power to the heat exchanger (MW) |
| $\eta_{rec}$ | Receiver efficiency |
| $P_{e,g}$ | Generated electrical power (MW) |
| $E_{e,g}$ | Generated electrical energy (kWh) |
| $dt$ | Time step, which equal one hour |
| $A_{land}$ | Land area (acres) |
| $A_{mirror}$ | Mirror area (m$^2$) |
| $CF$ | Capacity factor |
| $\eta_{field}$ | Field efficiency |

| | |
|---|---|
| $\eta_{sh\&bl}$ | Shading and blocking factor |
| $P_{th,tr}$ | Solar thermal power reflected by heliostats field (MW) |
| $A_{hs}$ | Heliostat area (m2) |
| $I(t)$ | The hourly output current, Amp. |
| $V(t)$ | The hourly output voltage, Volt |
| $A$ | The ideality factor for p-n junction |
| $T(t)$ | The hourly temperature, Kelvin |
| $KB$ | The Boltzman's constant in Joules per Kelvin, $1.38*10^{-23}$ J/k |
| $q$ | The charge of the electron in Coulombs, $1.6*10^{-19}$ C |
| $I_o(t)$ | The hourly reverse saturation current |
| $I_{ph}(t)$ | The hourly generated current of solar cells module |
| $T_r$ | The reference temperature |
| $E_{go}$ | The band-gap energy of the semiconductor used in solar cells module |
| $K_I$ | The short circuit current temperature coefficient |
| $I_{or}$ | The saturation current at $T_r$, Amp. |
| $\bar{H}_T(t)$ | The average hourly radiation on the tilted surface, kW/m$^2$ |
| $I_{sc}$ | PV cell short-circuit current at 25 °C and 100 mW/cm$^2$ |
| $P_{pv,out}(t)$ | The hourly output power of the solar cells module |
| $P_L$ | Load power |
| $N_{inv}$ | Total number of inverter |
| $P_{inv}$ | Rated power of inverter |
| $N_{pv}$ | Total number of modules |
| $V_{in}$ | The inverter rating voltage, V |
| $S_{in}$ | The inverter rating power, VA |
| $\eta_{in}$ | The inverter efficiency |
| $N_{series}$ | Number of series modules per each string |
| $V_c$ | The module voltage at maximum power point, Volt. |
| $N_{parallel}$ | Number of parallel strings per each subsystem |
| $P_{nom}$ | The nominal peak power for solar cells module, W |
| $N_3$ | Number of PV modules per subsystem |
| $N_{subsystem}$ | Number of subsystems. |
| $P_{g,total}(t)$ | Total generated power from CTR/PV system at any time of t |
| $\alpha$ | The system penetration ratio |
| $T_y$ | The number of hours through year |
| $CF_{hybrid}$ | The capacity factor of hybrid CTR/PV system |
| $T_{tr,out}$ | Tower receiver outlet temperature (c°) |
| $T_{tr,in}$ | Receiver inlet temperature (c°) |
| $\dot{m}_{CST}$ | Outlet mass flow rate from CST (kg/h) |
| $C_{p,HTF}$ | Specific heat of HTF (J/kg) |
| $\dot{Q}$ | Actual heat transfer rate (MW) |
| $U$ | Overall heat transfer coefficient (W/m$^2$·K) |
| $A$ | Heat transfer area respectively (m$^2$) |
| $\varepsilon$ | Heat transfer effectiveness |
| $Q_{max}$ | Maximum heat transfer rate between the two fluids (MW) |
| $\dot{m}_{HST,ref}$ | HTF mass flow rate at the reference full-load condition (kg/h) |

| | |
|---|---|
| $\dot{m}_{HST}$ | Outlet mass flow rate from the HST that equals to inlet mass flow rate to the heat exchanger (kg/h) |
| $T_{HTF,in}$ | HTF inlet temperature to the heat exchanger (c°) |
| $T_{steam,in}$ | Steam inlet temperature to the heat exchanger (c°) |
| $C_{min}$ | Minimum heat capacitance (W) |
| $\varepsilon_{ref}$ | Heat transfer effectiveness at the reference full-load condition |
| $C_r$ | Heat capacitance ratio |
| $h_4$ | Enthalpy after the pump (J/kg) |
| $h_3$ | Enthalpy before the pump (J/kg) |
| $P_4$ | Pressure after the pump (MPa) |
| $P_3$ | Pressure before the pump (kPa) |
| $\rho$ | Water density (kg/m$^3$) |
| $\eta_{pump}$ | Pump efficiency |
| $h_1$ | Actual turbine inlet enthalpy (J/kg) |
| $h_{2s}$ | Ideal turbine outlet enthalpy (J/kg) |
| $l_{tur}$ | Turbine specific work (J/kg) |
| $\eta_{tur}$ | Turbine efficiency |
| $T_1$ | Turbine inlet temperature (c°) |
| $P_1$ | Turbine inlet pressure (MPa) |
| $h_2$ | Actual turbine outlet enthalpy (J/kg) |
| $\dot{m}_{steam}$ | Steam mass flow rate (kg/s) |
| $T_{steam,in}$ | Turbine inlet steam temperature (c°) |
| $\eta_{gen}$ | Generator efficiency |
| $P_{ele}$ | Electrical output power of CTRPP (MW) |
| $M_{HST}$ | Mass inside HST (kg) |
| $M_{CST}$ | Mass inside CST (kg) |
| $x_0$ | Initial condition for the mass (kg) inside the CST and HST |
| $t_0$ | Lower limit of the two tanks (kg) |
| $t_u$ | Upper limit saturation of the two tanks (kg) |
| $E_{stored}$ | Stored thermal energy (MWh) |
| $T_{HST,salt}$ | Salt temperature in the HST (c°) |
| $T_{CST,salt}$ | Design salt temperature in the CST (c°) |
| $M_{run}$ | A predetermined value of HTF mass inside HST at which the discharge starts (kg) |
| $M_{st}$ | A minimum value of HTF mass inside HST at which the discharge stop (kg) |
| $T_{HST,out}$ | HST outlet temperature (c°) |
| $T_{st}$ | Temperature setting (c°) |
| $P_{sol\text{-}min}$ | Minimum value of solar thermal energy |
| $S_0$ | Sum of bias value and weighted inputs created by neurons |
| $Z_i$ | ANN inputs data |
| $W_{ij}$ | Neurons weights |
| $b_j$ | Neuron bias |
| $n$ | Net input argument |
| $Y$ | ANN output |

| | |
|---|---|
| $F$ | Transfer function |
| $Z$ | Input vector |
| $H_n$ | Neurons numbers in hidden layers |
| $N_T$ | Number of training data |
| $N_I$ | Number of neurons in input layers |
| $N_O$ | Number of neurons in output layers |
| $\sigma_j$ | Width of *j*th neuron (spread factor) |
| $j$ | Number of neuron in hidden layer |
| $\mu_j$ | RBF centre unit |
| $S_s$ | S-summation neuron in summation layer |
| $G(z, zi)$ | Gaussian function |
| $z_j$ | Represent the *j*th element of $z$ |
| $z_{ij}$ | Represent the *j*th element of $z_i$ |
| $Z_{actual}$ | Actual value of the target |
| $Z_{predicted}$ | Predicted value by the ANN model |
| $N_S$ | Total number of samples |
| $R^2$ | Coefficient of determination |
| $R$ | Correlation coefficient |
| $Z_1$ | The first input of ANN model, which represents $P_{th,tr}$ at the tower receiver |
| $Z_2$ | The second input of ANN model, which represents $T_{in,tr}$ of the HTF mass flow rate |
| $T_{tr,in}$ | Inlet receiver temperature (c°) |
| $T_{tr,out}$ | Outlet receiver temperature (c°) |
| $T_{normal}$ | The normalized value ofthe variable $T_i$ |
| $\text{High}_{value}$ | The high value that equal +1 |
| $\text{Low}_{value}$ | The low value that equal −1 |
| $Ti$ | Variable which represents the target |
| $T_{max}$ | Represents the values of the maximum of variable $T_i$ |
| $T_{min}$ | Represents the values of the minimum of variable $T_i$ |
| $K_{invest}$ | Total plant investment ($) |
| $K_{fuel}$ | Annual fuel cost ($) (CSP plants have virtually zero fuel costs, i.e., $K_{fuel} = 0$) |
| $K_{O\&M}$ | Aannual operation and maintenance ($) |
| $k_d$ | Discount rate (%) |
| $m$ | Analysis period (year) |
| $k_{insur}$ | Annual insurance rate (%) |
| $F_C$ | Fixed cost by capacity ($/kW-year) |
| $L_C$ | Land cost per unit area ($/m$^2$) |
| $M_C$ | Mirror cost per unit area ($/m$^2$) |
| $V_C$ | Variable cost by generation ($/MWh) |
| $h_{rec}$ | Receiver height (m) |
| $h_{hel}$ | Heliostat height (m) |
| $W_{hel}$ | Heliostat width (m) |
| $A_{rec}$ | Receiver area (m$^2$) |
| $A_{ref,rec}$ | Receiver reference area (m$^2$) |

| | |
|---|---|
| $C_{TS}$ | Cost of thermal storage per $kWh_{th}$ ($/$kWh_{th}$) |
| $E_{tes}$ | Designed stored thermal energy (MWh) |
| $T_S$ | Storage hours (hour) |
| $\eta_{ts}$ | Storage system efficiency |
| $k_{con}$ | Constant value as a percent of subtotal cost |
| $k_{EPC}$ | Constant value as a percent of the direct cost |
| $P_{min,conv}$ | Minimum generation constraint (MW) |
| $P_{CSP}$ | Concentrated solar power system Output (MW) |
| $P_{PV}$ | PV system output power (MW) |
| $P_{g,conv}$ | Conventional generators output power (MW) |
| $P_{PV,Dissip}$ | Curtailed PV power (MW) |
| $P_{ch}$ | Charge power of the storage system |
| $P_{disch}$ | Discharge power of the storage system |

# List of Figures

# List of Tables

# Chapter 1
# Introduction and Literature Review

**Abstract** As the world's supply of fossil fuels shrinks, there is a great need for clean and affordable renewable energy sources (RES) in order to meet growing energy demands. Furthermore, the conventional plants based on fossil fuel have serious environmental and financial problems, and therefore, the dependency of the distribution networks on the RES such as solar power systems for generating electrical power is significantly promoted. In the past few decades, solar energy systems have been received great attention as an important type of RES. Nowadays, solar energy sources constitute appropriate commercial options for small and large power plants. The two mainstream categories of solar energy systems utilized for this purpose are concentrated solar power (CSP) and photovoltaic (PV). This chapter presents a brief introduction about the role, important need, and advantages of renewable energies for today and the future, especially solar energy such as PV and CSP systems. In addition, it introduces a survey for all types of CSP technologies. As well as, it presents a literature review for the LCOE and cost reduction of CSP and PV systems, CSP modeling, and the application of ANN technologies in various SF systems. Further, it presents the problem definition, objectives, and outlines of this thesis.

**Keywords** Renewable energy sources (RES) · Concentrated solar power (CSP) · Central receiver tower (CTR) · Parabolic trough (PT) · Parabolic dish (PD) · Linear fresnel (LF) · Photovoltaic (PV) · Thermal energy storage (TES) · Cost reduction · CSP models

## 1.1 Introduction

As the world's supply of fossil fuels shrinks, there is a great need for clean and affordable renewable energy sources (RES) in order to meet growing energy demands. In the past few decades, solar energy systems have been received great attention as an important type of RES. The advantages of solar energy (e.g., clean, abundant, a source with a free cost, and environmentally friendly energy solutions) make it one of the most promising technology in the world. In general, RES technologies are considered as an alternative solution to reduce $CO_2$ emission, so they could be a key technology for mitigating climate change [1, 2]. The conventional plants based


I. Moukhtar et al., *Solar Energy*, Power Systems,
https://doi.org/10.1007/978-3-030-61307-5_1

on fossil fuel have serious environmental and finical problems, and therefore, the dependency of the distribution networks on the solar power systems for generating electrical power is significantly promoted [3]. Also, solar power is characterized by a daily and seasonal pattern, where peak output usually occurs in the middle of the day and in the summer, so it is quite well correlated with the hours of the high demand of many electric power systems [4].

Nowadays, solar energy sources constitute appropriate commercial options for small and large power plants [3]. There are two mainstream categories of solar energy systems utilized for this purpose: concentrated solar power (CSP) and photovoltaic (PV) configurations. A block diagram shows the different types of solar energy systems that convert the solar resource into a useful form of energy is given in Fig. 1.1. The downside of CSP and PV technologies without storage systems is that they have non-controllable variability, partial unpredictability and locational dependency. Where the large-scale penetration of intermittent renewables is expected to have profound implications on many aspects of power systems planning, operation and control, as well as on the corresponding regulation [5, 6]. So, with large shares of CSP and PV technologies predicted in the future, steps should need to be taken to ensure a continued and reliable supply of electricity.

Many feasible solutions have been done in the field of storage systems to solve the dispatchability issue of these technologies. Regarding CSP technologies, considerable efforts are put in order to permit these technologies to become truly dispatchable and penetrate actively in the electrical grid. At this point, CSP with thermal energy storage (TES) plays a key role in power generation, particularly during cloudy weather periods and after sunset or in early morning when power demand steps up.

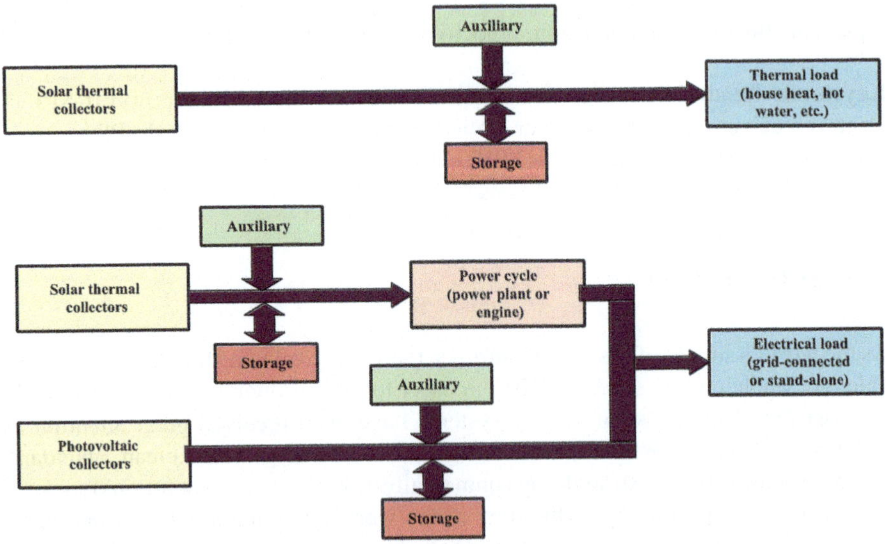


**Fig. 1.1** Diagram of a basic solar energy conversion systems

In large-scale power plants (>50 MW), the competitiveness in terms of cost and efficiency of storing energy can turn in favour of CSP plants compared to PV plants [6, 7]. Another potentially important benefit of CSP systems integrating TES, along with dispatchability, is their ability to provide system flexibility: this feature might enable higher overall penetration of other variable generation technologies such as those based on PV cells and wind turbines [7]. More importantly, the International Energy Agency (IEA) reported that the expected contribution of CSP technologies will cover between 10–11.3% of the electricity production in the world by 2050 [8]. The CSP technologies revealed advantageous characteristics over classical energy resources such as:

- The inherent flexibility of CSP/TES plant that provides the overall system flexibility and enhances energy security; offer higher ramp rates and ranges than large thermal plants currently used to meet a large fraction of electric demand [9].
- The dispatchability of CSP with TES can enable higher overall penetration of solar energy in two ways. The first is providing solar-generated electricity during periods of cloudy weather or at night. However a potentially important, and less well analyzed benefit of CSP is its ability to provide system flexibility, enabling greater penetration of PV (and other variable generation sources such as wind) than if deployed without CSP [9].
- Less cost of TES especially at large scale of CSP compared with other technologies that utilize electrical storage forms [10].
- With concentrating solar power systems, there are no complicated silicon manufacturing processes, as in the case of PV systems; no deep holes to drill, as in the case of geothermal systems; and no turbine housings that need to be kept greased at high elevations from the ground, as in wind-power systems [11].
- Their construction resembles traditional power plants (i.e., uses many of the same equipment and technologies) and unlike PV technologies, CSP plants can provide most needed ancillary services [6].
- Compared to other energy sources, solar energy output is generally more predictable due to low forecast errors on clear days, and the ability to use satellite data to monitor the direction and speed of approaching clouds [4].

The design of all CSP power plants is complex and is usually accomplished using computer modeling. There is a number of computer programs available for the design of CSP plants. System Advisor Model (SAM) is one of the famous programs that used to simulate the output power of renewable energy such as PV, CSP and wind energy. SAM was developed by National Renewable Energy Laboratory (NREL) under contract with the US department of energy and is available for free [12].

This book presents computer programs for the optimal design of central tower receiver (CTR) and PV systems. The proposed programs were implemented to find the optimal design parameters such as the number of heliostats field, land area, number of PV modules and inverter units. The computer programs can completely design the central tower receiver power plant and PV systems interconnected with an electric grid and determine the annual output power. Therefore, it can estimate the monthly energy excess, monthly energy shortage, and yearly purchase or selling energy to/or from the electric grid. Additionally, this program is designed to calculate the minimum energy cost ($/kWh). Furthermore, this work presents a simple and accurate mathematical model for CTR/TES. Furthermore, a facile controllable scheme is adopted using artificial neural network (ANN) technique for modeling this plant, which is one of the important types of CSP technologies. Significantly, the incorporated ANN method simplifies the calculations processing and solves the complexity issue of the CTR model due to its functional ability to fix the nonlinearity. ANN technique is used to evaluate the mass flow rate from cold storage tank (CST) to the receiver based on the value of solar thermal power at the tower receiver. Therefore, it can adjust or control the receiver outlet temperature along the operating time regardless of the variations in the direct solar radiation and receiver inlet temperature. Also, in this book, CTR/TES has been modeled for partial load conditions in order to accommodate the required electrical power and intermittent solar energy, especially during the periods of cloudy weather and after sunset. As well, the algorithm of electrical generation methodology was modified for regulating CTR/TES output according to the hot storage tank (HST) conditions.

Finally, owing to the negative impact of high penetration of PV system, CSP/TES has been aggregated with PV system in order to accommodate the required electrical power during the higher and lower solar energy at all timescales. More importantly, this work analyzes the impacts of CSP with TES on the system reliability and penetration level of PV system without electrical batteries.

## 1.2 CSP Technologies

The sun's total power output is $3.8 \times 10^{20}$ MW, which is equal to 63 MW/m$^2$ of the sun's surface. This power radiates outward in all directions [11]. The earth receives only a tiny fraction of the total radiation emitted, equal to $1.7*10^{14}$ kW; however, even with this small fraction, it is estimated that 84 min of solar radiation falling on earth is equal to the world energy demand for one year [11]. So, the solar radiation that reaches the Earth's surface is an important RES. Where, this radiation can be converted into thermal energy through CSP systems that can be utilized for power generation or it can be transformed directly into electricity using PV systems. The CSP technologies use mirrors to concentrate the sunlight for creating a high energy density and temperature level that drives steam turbines traditionally powered from conventional fossil fuels. These types of systems can only work with direct solar radiation, so they are more likely to be used in areas where there are few clouds because otherwise PV technologies would fit better [13–15]. Because of the large area required for CSP plants, these are usually located on non-fertile ground, such as deserts [11]. According to the trans-Mediterranean renewable energy corporation (TREC), each square kilometer of the desert receives solar energy equivalent to 1.5 million barrels of oil. It has also been estimated that, if an area of desert measuring 65,000 km$^2$, which is less than 1% of the Sahara desert, were covered with CSP plants, it could produce electricity equal to the year 2000 world electricity consumption [11,

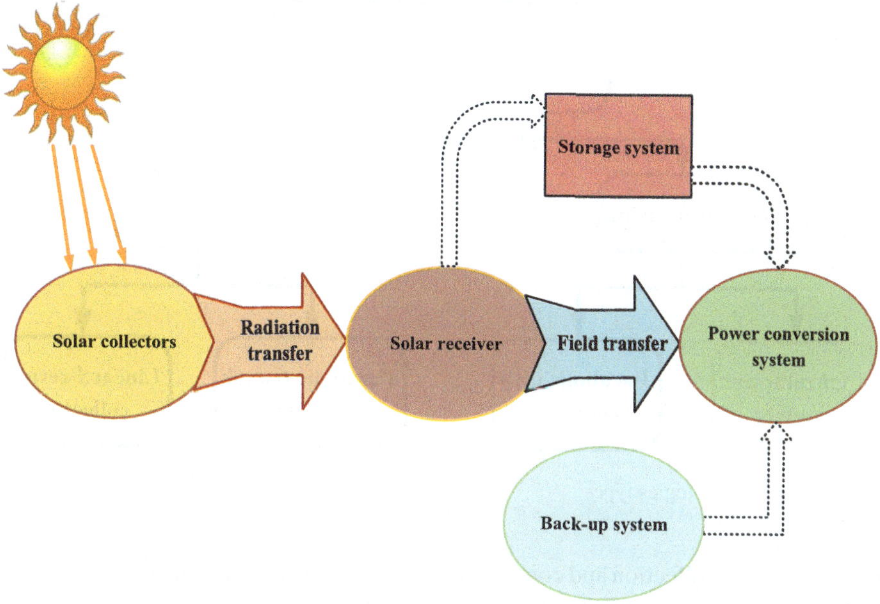


**Fig. 1.2** Schematic diagram of solar thermal energy conversion system using CSP technologies

16]. One fifth of this area could produce the current electricity consumption of the European Union. Similar studies in the United States (US) predict that the solar resource in southwestern states could produce about 7000 GW with CSP, which is about seven times the current total US electric capacity [11, 17].

The main three parts of CSP plant components are solar energy collection and concentration (i.e., heliostats field and receiver system), heat transfer fluid (HTF) and storage system (i.e., CST, hot storage tank 'HST', and molten salt), and power block (PB) system (i.e., steam generator 'SG', turbine, and electric generator) [18]. Figure 1.2 shows the basic schematic of CSP plant for converting solar energy into electrical power. In the CSP systems, the incoming radiation is tracked by large mirror fields, which concentrate the energy towards absorbers (receiver). The receiver receives the concentrated radiation and transfers it thermally to the working medium, which may be molten salt or oil or water. Some of these systems also incorporate TES, which stores the extra unused heated molten salt for future use or during cloudy weather and nighttime. When the electrical energy is required, the heated molten salt is pumped through a heat exchanger immediately to produce the required superheated steam which then drives the classical PB (i.e., Rankine cycle turbine/generator system) to generate the electricity [19, 20]. In a hybrid plant, back-up and/or storage systems are added to enhance the performance and increase the capacity factor (CF) [21, 22]. It is clear from Fig. 1.2 [22] that the three parts of CSP plant components are follows:

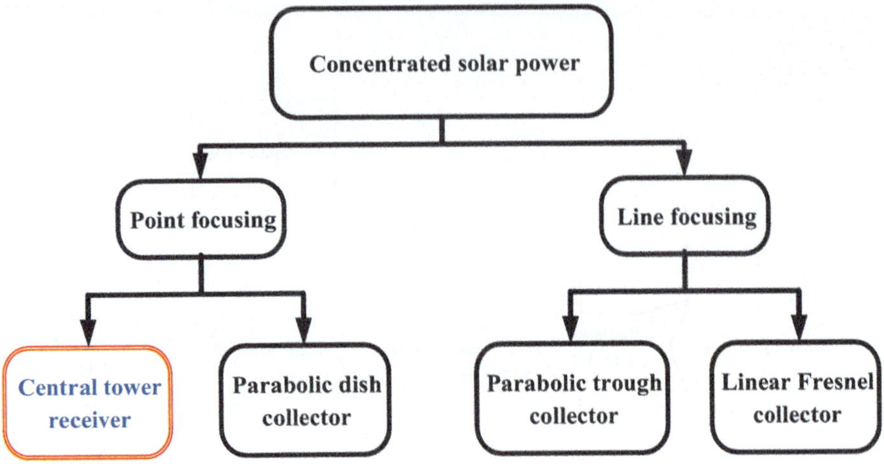


**Fig. 1.3** CSP technologies types

- Solar energy collection and concentration; i.e., heliostats field and receiver system.
- HTF and storage systems; i.e., CST, HST and molten salt.
- PB system; i.e., SG, turbine, and electric generator.

Based on the method type of collecting and concentrating solar radiation, CSP systems can be classified into groups as shown in Fig. 1.3 [23].

- One that focus the sun rays to a point such as CTR and parabolic dish (PD).
- Other those focus the sun rays to a line such as parabolic trough (PT), Linear Fresnel (LF) reflector.

This work will focus on CTR technology as one of the most important types of CSP systems, where it has greatly improved in the last few decades and continues to take more attention as a suitable system for the large solar thermal plants.

### *1.2.1 World Current Status of CSP Market*

After twenty After twenty years of operation in the solar electric generating system plants in California, the world-wide market growth of renewable energies has given CSP technology a new prospective in countries with high direct radiation. Starting in the Spanish and US electricity markets, many projects are now under development and under construction [24]. From European and Mediterranean perspective and analyzing direct normal irradiation (DNI) potential, two promise countries are easily identified in Europe for the development of thermal solar energy plants, namely as Spain and Portugal. In North Africa region there are Morocco, Algeria, Libya

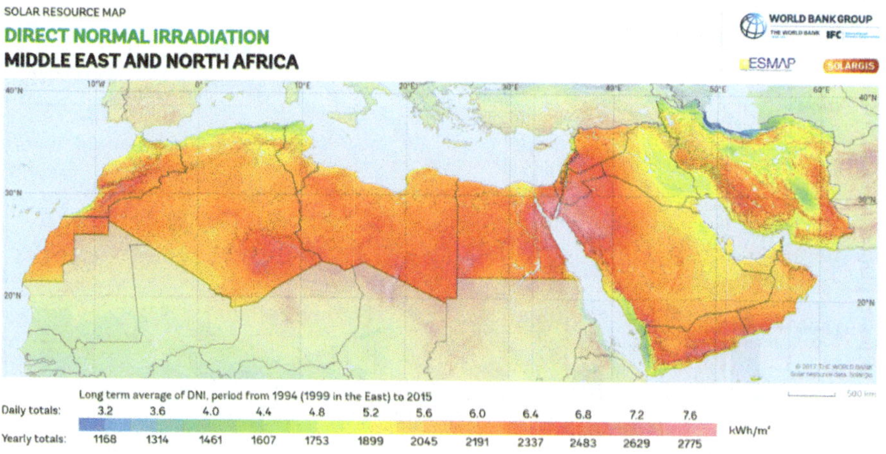


**Fig. 1.4** DNI potential for MENA regions [25]

and Egypt and in the Middle-East there are Saudi Arabia and Jordan as potential countries to install these solar-based systems as shown in Fig. 1.4 [18, 25]. Among all these countries, Spain is the number one in CSP plant investment and solar energy production; the leader solar thermal energy world producer and pioneer in Europe and Mediterranean regions. For political and economic reasons, many Middle East and North Africa (MENA) countries are far behind Spain. On the other hand, USA is the world leader due its strong investments in CSP technology and other new projects [18].

Table 1.1 shows the size of different CSP technologies according to the project status and lists the current CSP projects in the world market [24]. While Fig. 1.5 presents a global view of CSP plant planned, operating and under constructing. It can be seen the higher power capacity operating CSP plants in Spain, with two times the power capacity of the USA. However, due to the investments made in the USA and the current construction of new CSP plants, they will quickly produce more power than any other country in the world. As exhibited in Fig. 1.6, other countries in MENA are developing their first projects; if implementation is successful, further projects are expected in all of these countries [18, 24].

**Table 1.1** Current CSP projects in the world mark [18, 24]

| Parameters | Operational (MW) | Under construction (MW) | Planning phase (MW) | Total (MW) |
|---|---|---|---|---|
| Central tower | 44 | 17 | 1603 | 1664 |
| Parabolic trough | 778 | 1400 | 8144 | 10,322 |
| Linear fresnel | 9 | 30 | 134 | 173 |
| Parabolic dish | 2 | 1 | 2247 | 2250 |
| Total | 833 | 1448 | 12,128 | 14,409 |

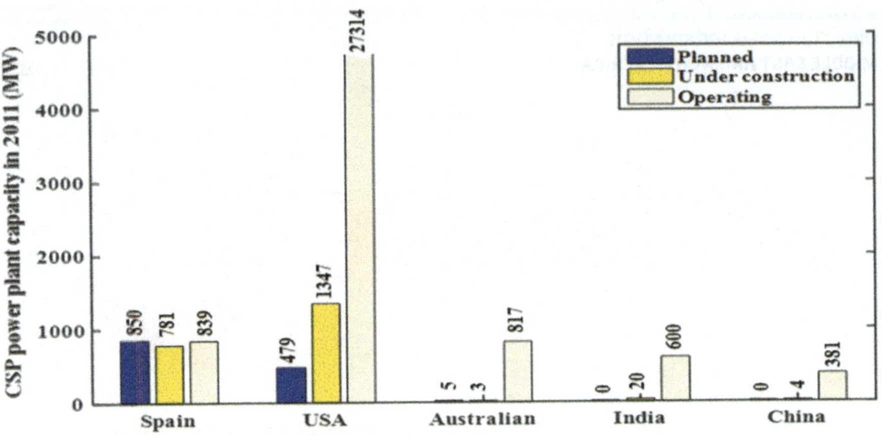


**Fig. 1.5** Worldwide distribution of operating, under construction, and planed CSP plants

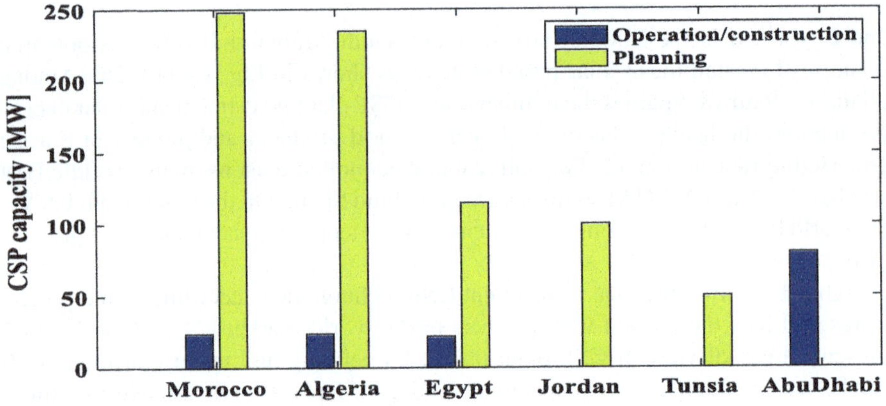


**Fig. 1.6** Capacities of CSP projects in MENA

### *1.2.2 CTR Technology*

Recently, the central tower receiver power plant draws extensive attention as a promising candidate for the large solar thermal plants. This is mainly due to the expected performance improvements and cost reductions associated with technology innovations of the three main subsystems, i.e., the heliostat, the receiver and the PB within the near future [26]. The central tower receiver power plant is known as one of the least expensive power plants to produce solar electricity on a large scale, also it has the best performance [19]. Additionally, this technology exhibited high operating temperature, which leads to a reduction in the cost of TES systems and it is able to cover the demand as modern power conversion systems [10]. The central receiver concentrator is suitable for thermal electric power production in the range of 10–1000 MW due to its high operating temperature [12].

As displayed in Fig. 1.7a, the central receiver systems use a large number of computer-assisted mirrors, heliostats field or solar field (SF), oriented around a tower for tracking the sun by two axes mechanisms. The heliostats fields reflect the sunlight and concentrate it towards a fixed central receiver situated atop the tower [10]. The receiver absorbs the reflected solar radiation by heliostats field and converts it into heat at high temperature levels. A power conversion system is used to shift thermal energy into electricity in the same way as conventional power plants. Higher concentration ratios are achieved by this system compared to linear focusing systems and this allows thermal receivers to operate at higher temperatures about 1000 °C, with reduced losses. The surrounding heliostat field has a significant cost contribution in central tower receiver power plant, almost 50% of total cost and 40–47% of the total losses is assigned to the heliostat field. Therefore, the design and

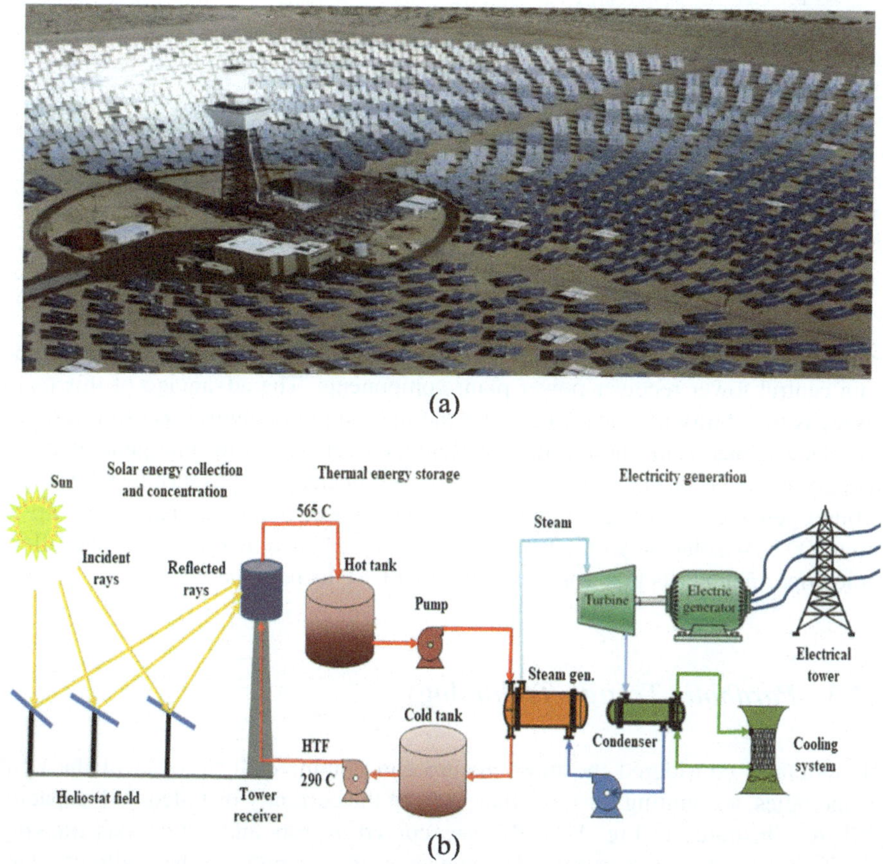


**Fig. 1.7** **a** Typical CTR plant in California [27] and **b** Schematic diagram of CTR components

Table 1.2 Main specifications of the four CSP technologies [10, 26]

| Parameters | PT | CTR | LF | PD |
|---|---|---|---|---|
| Solar collector | Line focus | Point focus | Line focus | Point focus |
| Solar receiver | Mobile | Fixed | Fixed | Mobile |
| Operating temp. °C | 350–550 | 250–565 | 390 | 550–750 |
| Plant peak efficiency % | 14–20 | 23–35 | 18 | 30 |
| Typical capacity MW | 10–300 | 10–200 | 10–200 | 0.01–0.025 |
| Grid stability | Medium | High (large TES) | Medium | Low |
| Storage systems | available | available | Possible | Possible |
| Relative cost | Low | High | Very high | Very low |
| Tracking system | One-axis | Two-axis | One-axis | Two-axis |
| Technology risk | Low | Medium | Medium | Medium |
| Future improvements | Limited | Very significant | Significant | not proven |
| Efficiency with improvements % | 18 | 25–28 | 12 | 30 |
| Efficiency rise after improvements % | 20 | 40–65 | 25 | 25 |
| Power conversion cycle | Superheated Rankine steam cycle | Superheated Rankine steam cycle | Saturated Rankine steam cycle | Stirling cycle |

optimization of heliostat field layout are very important [26]. Figure 1.7b shows the main central tower receiver power plant components. The advantage of this technology is the ability to convert the solar thermal output of central receiver systems into electric energy in highly efficient Rankine-cycle/steam turbine generators, in Brayton-cycle/gas turbine generators, and in combined cycle (gas turbine with steam turbine) generators [26]. Grid-connected tower power plants are applicable up to about 200 MW solar-only unit capacity [19]. As explained in Table 1.2, CTR offers some important merits that don't exist in other CSP technologies.

### 1.2.3 Parabolic Trough Technology

PT system is considered the most mature and commercially proven of the CSP technologies, accounting for more than 90% of the currently installed CSP capacity [28]. As illustrated in Fig. 1.8a, PT uses curved mirrors and single-axis tracking to follow the sun throughout the day, concentrating sunlight on thermally efficient

(a)

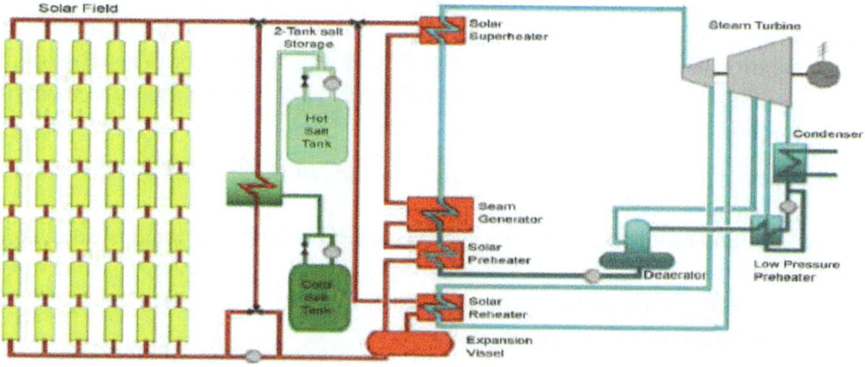


(b)

**Fig. 1.8** **a** PT collector and **b** PT power plant schematic diagram [18]

receiver tubes or heat collection elements. The collector can be oriented in an east–west direction, tracking the sun from north to south, or in a north–south direction, tracking the sun from east to west.

The advantages of the former tracking mode is that little collector adjustment is required during the day and the full aperture always faces the sun at noon but the collector performance during the early and late hours of the day is greatly reduced, due to large incidence angles (cosine loss). North–south oriented troughs have their highest cosine loss at noon and the lowest in the mornings and evenings, when the sun is due east or due west. Over a period of one year, a horizontal north–south through field usually collects slightly more energy than a horizontal east–west one. However, the north–south field collects a lot of energy in summer and much less in winter. The east–west field collects more energy in winter than a north–south field and less in summer, providing a more constant annual output. Therefore, the choice of orientation usually depends on the application and whether more energy is needed during summer or winter [11]. The receiver comprises the absorber tube (usually metal) inside an evacuated glass envelope. The absorber tube is generally coated stainless steel tube, with a spectrally selective coating that absorbs the solar (short wave) irradiation well, but emits very little infrared (long wave) radiation. This helps to reduce heat loss. Evacuated glass tubes are used because they help to reduce heat losses [29]. A HTF, typically synthetic oil; molten salt; or steam, circulates in the tubes absorbing the sun's heat before passing through multiple heat exchangers to produce steam. Most existing PT systems use synthetic oils as a HTF, which are stable up to 400 °C. The use of molten salt at 540 °C for either heat transfer or storage purposes is under demonstration [10]. The steam spins a conventional steam cycle turbine to generate electricity or it is integrated into a combined steam and gas turbine cycle when used in hybrid configurations. Utility-scale collector fields are made up of many parallel rows of troughs connected by receiver tubes in series as presented in Fig. 1.8b. In utility settings, solar trough power plants have shown consistent performance when connected to the electric grid. Improved operating flexibility and dispatchability has been achieved through integration with hybrid fossil systems as well as through demonstrated TES capabilities [27]. The PT systems have the following shortages [19]:

- The upper process temperature is currently limited by the heat transfer thermal oil to 400 °C.
- Low-cost and efficient energy storage systems have not been demonstrated up to now.
- The heat transfer thermal oil adds extra costs of investment and of operating and maintenance.
- Some absorber tubes are still subjected to early degradation; reasons are the risk of breakage of absorber envelope glass tubes with loss of vacuum insulation and degradation of the absorber tube selective coating.
- High winds may break mirror reflectors at field corners.
- Direct steam generation trough technology is still in a developmental stage.

### *1.2.4 Parabolic Dish Technology*

Dish systems use a dish-shaped concentrator (like a satellite dish) to focus the sun's rays onto a receiver mounted at the focal point as demonstrated in Fig. 1.9. In the receiver a heat-transfer medium takes over the solar energy and transfers it to the power conversion system, which may be mounted in one unit together with the receiver (e.g. receiver/Stirling engine generator unit) or at the ground. The receiver may be a Stirling engine and generator (dish/engine systems) or it may be a type of PV panel that has been designed to withstand high temperatures (concentrated PV systems) [17]. Due to its ideal optical parabolic configuration and its two axes control for tracking the sun, dish collectors achieve the highest solar flux concentration, and therefore the highest performance of all concentrator types in terms of peak

**Fig. 1.9** PD system [11]

solar concentration and of system efficiency. These collector systems are restricted to unit capacities of some 10 kW for geometrical and physical reasons. The dish technology is applicable to off-the-grid power generation, i.e. at remote places or at island situations. Dish systems may optionally be arranged in large dish arrays in order to accumulate the power output from the kW capacity up to the MW range [19]. PD systems can achieve temperatures in excess of 1500 C. Because the receivers are distributed throughout a collector field, like PT systems, PD systems are often called distributed receiver systems [11].

The main advantages of PD systems include high efficiency (i.e. up to 30%) and modularity (i.e. 5–50 kW), which is suitable for distributed generation. Unlike other CSP options, PD systems do not need cooling systems for the exhaust heat. This makes PDs suitable for use in water-constrained regions, though at relatively high electricity generation costs compared to other CSP options [28]. PD systems have the following shortages [19]:

- The electricity output of single dish/Stirling unit is limited to small ratings of e.g. 25 kW due to geometric and physic reasons (exception: Australian big dish designed for use of a 50 kW steam engine or turbine generator).
- No adequate energy storage system is applicable or available.
- The establishment of industrial large volume production of dish components and stirling engines is needed for entry into appropriate market segments.
- Large-scale deployment has not yet occurred.

### 1.2.5 Linear Fresnel Reflector Technology

The ‘Fresnel mirror’ type of CSP system is broadly similar to PT systems except it uses long flat mirrors at different angles as shown in Fig. 1.10a. These flat mirrors have the effect of focusing sunlight on one or more tubes, which are mounted above the mirrors. As with PT systems, the mirrors change their orientation throughout the day so that sunlight is always concentrated on the heat-collecting pipe. The receiver is stationary and does not move with the mirrors as in the CSP trough systems, so it does not require rotating couplings between the receivers and the field header piping, thus providing additional design flexibility [27]. LF plant schematic diagram is shown in Fig. 1.10b.

The main advantages of LF systems compared to PT systems are as the following [10, 17]:

(a)

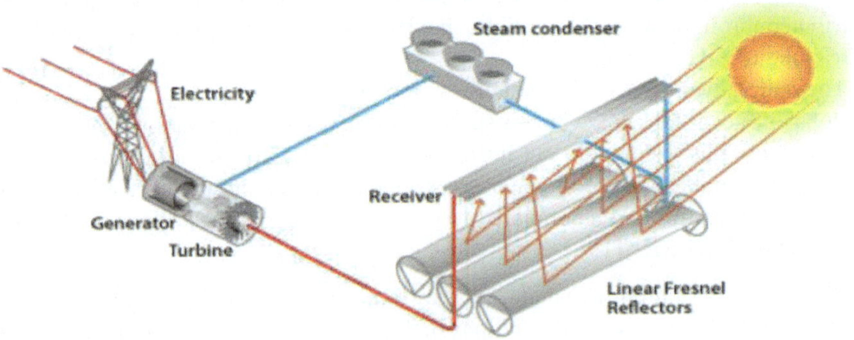


(b)

**Fig. 1.10** **a** LF system [18] and **b** LF plant schematic diagram [27]

- LF system uses flat or elastically curved reflectors, which are cheaper than parabolic glass reflectors.
- LF reflector are mounted close to the ground, thus minimizing structural requirements.
- The wind loads on LF collectors are smaller, resulting in better structural stability, reduced optical losses and less mirror-glass breakage.

These advantages need to be balanced against the fact that the optical efficiency of LF collector solar fields is lower than that of PT collector solar fields due to the geometric properties of LF collectors. The problem is that the receiver is fixed and in the morning and afternoon cosine losses are high compared to PT collector. Despite these drawbacks, the relative simplicity of the LF collector system means that it may be cheaper to manufacture and install than PT-CSP plants. However, it remains to be seen if costs per kWh are lower. Additionally, given that LF collectors are generally proposed to use direct steam generation (i.e., water is the typical thermal fluid which flows through the SF tubes to generate the steam, eliminating the need for costly heat exchangers). Therefore, adding TES is likely to be more expensive [10]. A brief comparison between families of CSP technologies is illustrated in Table 1.2, which describes the main specifications of these technologies [10, 26].

## 1.3 Thermal Energy Storage

All forms of energy are either potential energy (e.g. chemical or gravitational), kinetic energy, electrical energy or thermal energy, and all these forms of energy could be stored with an appropriate method and technology. As shown in Fig. 1.11, a large variety of energy storage systems are under development [30]. TES will be discussed in this book, because it is the best method to be applied in solar power plants.

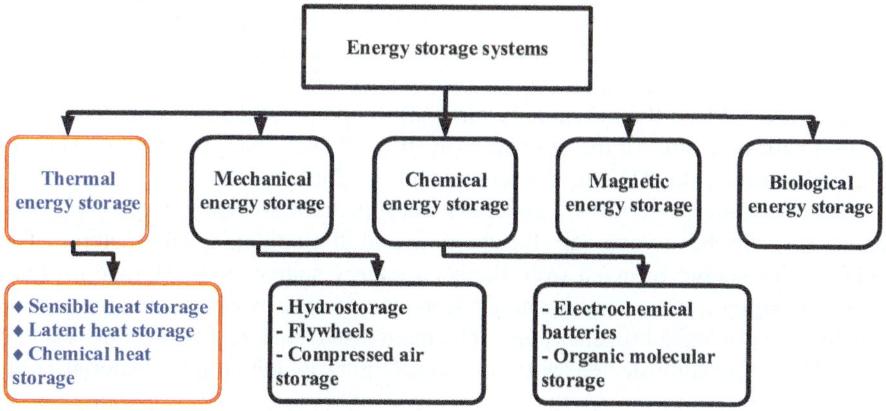


**Fig. 1.11** Classification of energy storage systems

TES is a storage media that stores the heat from SF during charge periods (daylight periods) and releases it during discharge process during overcast or nighttime periods. Incorporating TES into CSP plants increases the number of operation hours, system energy efficiency, and PB utilization and reduces the times of mismatch between energy demand and energy supply by the sun; this leads to make the integration of electric grid easier [18]. However, its cost is almost 25% of the total cost, which lead to rise the power generation cost from the solar power plants. TES systems can be divided into three categories comprising: sensible heat; the energy absorbed or released by a material due to increase or reduction of its temperature, respectively, latent heat; storing energy by the transition heat from a solid to liquid state, and thermo-chemical; it is essential that chemical reactions involved are completely reversible [18, 31]. Sensible and latent thermal energy systems can keep the stored energy efficiently for a long time, while the third type retains its energy for short-periods. Furthermore, there are several types of thermal energy fluids that can be used as a HTF, depending on CSP technology type and storage system required, such as molten salts, oils, hydrogen, helium, and steam. Indeed, molten salts are the only commercial choice to store the thermal energy because of they provide many merits compared to other fluids. The possibility of incorporating thermal storage systems into CSP plants allows increasing the energy efficiency of the system, correcting the deviations between the generation and consumption, and making the electric network integration easier [18]. In what concerns the concept of the thermal storage process in solar power plants, there are active and passive systems.

The active storage system is mainly characterized by the forced convection heat transfer into the storage material. The storage medium itself circulates through a heat exchanger (this heat exchanger can also be a solar receiver or a SG). This system uses one or two tanks as storage media. Active systems are subdivided into direct and indirect systems [32]. One of the active direct systems is the two tanks direct system as shown in Fig. 1.7b, where the HTF is directly stored in a HST in order to use it during cloudy periods or nights. The cooled HTF is pumped to the other tank (i.e., CST) where it remains waiting to be heated one more time [32]. On the other side, the active indirect systems have two different types of HTF which one is used for circulating in the SF system and other for the storage medium as revealed in Fig. 1.8b. On other word, in that type the energy is not stored directly by the HTF but by a second heat fluid (generally oil) [32]. In the present work, an active direct system with two tanks of molten solar salt (60% $NaNO_3$ and 40% $KNO_3$) is used; one tank for the cold HTF and the second tank for the hot HTF.

Passive storage systems are generally dual medium storage systems: the HTF passes through the storage only for charging and discharging a solid material. The HTF carries energy received from the solar energy source to the storage medium during charging, and receives energy from the storage when discharging (these systems are also called regenerators). The main disadvantage of regenerators is that the HTF temperature decreases during discharging as the storage material cools

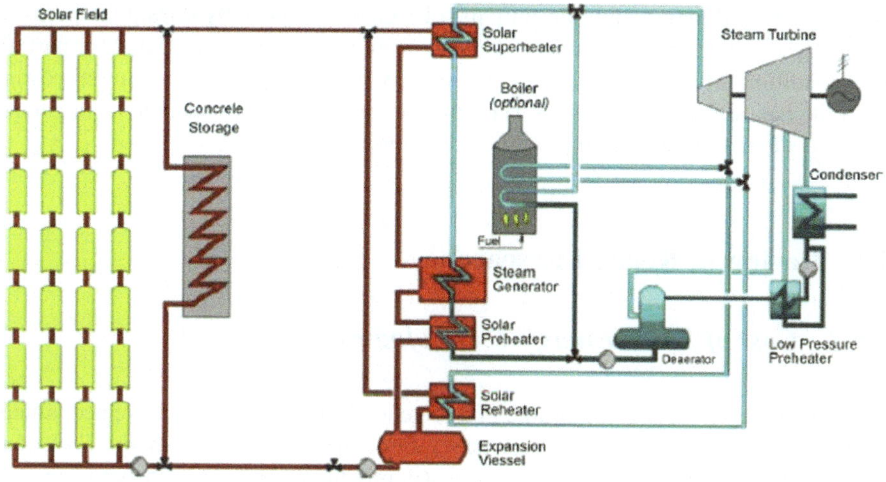


**Fig. 1.12** Scheme of a PT power plant with a concrete storage system [18]

down. As shown in Fig. 1.12, in the concrete storage case for example, the solar energy of the SF is transferred from the HTF to the solid storage material system. The storage material contains a tube heat exchanger to transfer the thermal energy from the HTF to the storage material [18].

## 1.4 Power Block and Steam Generator

Nowadays, there are many efforts to take advantage of the waste heat to produce electricity. The organic Rankine cycle (ORC) systems are considered as promising solutions for power generation from lower and medium temperature heat sources. Hence, ORC systems are usable alternatives to the conventional Rankine steam power plants at temperatures range from 100 to 350 °C. The operating cycle of ORC is the same principle of conventional steam Rankine cycle; however, the ORC uses an organic working fluid. The selection of the organic working fluid of ORC system is a critical challenge and depends on the temperature of the available heat source. Moreover, the turbine design is a key parameter of ORC system because of the high expansion ratios and small volumetric flow rates. The ORC systems displayed revealed advantageous features of high thermal efficiency, low cost, efficient solution to convert the heat to electric power. Therefore, these systems are considered as reliable technologies for renewable energy field such as solar energy, biomass and geothermal power plant [33, 34]. However, in this work, the used power cycle in the model of the adopted CTR plant is a conventional steam Rankine cycle of 40 MW. Furthermore, the shell and tube heat exchanger is a common type of heat exchanger that used in the solar power plants [35]. Also, in order to accommodate the load demand nature and intermittent solar energy, especially during the periods of cloudy weather and after sunset, the heat exchanger could be modeled for partial load conditions. Therefore, the outlet steam flow rate depends on the required power from the heat exchanger, thus the HTF inlet mass flow rate is variable to maintain constant steam temperature through the operation cycle.

## 1.5 Review of Related Researches

### *1.5.1 Cost of Solar Energy Technologies*

In every year since 2011, renewable power generation technologies have accounted for half or more of total new power generation capacity added globally. For solar electricity generating technologies to be cost competitive at a large scale with conventionally generated electricity, cost reductions are needed for both CSP and PV systems [36]. A virtuous circle of support policies around the world have become increasingly effective, resulting in increased deployment, technological improvements and cost reductions [37]. In 2011, the U.S. Department of Energy (DOE) established solar cost targets that corresponded to reducing CSP and PV prices by approximately 75% in order to achieve levelized cost of electricity (LCOE) of $0.06 per kWh for both utility-scale PV and high-capacity factor CSP-TES systems in 2020. To examine the implications of achieving this goal, DOE's solar energy technologies office published the SunShot Vision Study, which found that achieving the 2020 cost targets could result in significant solar penetration by 2030 [36].

#### 1.5.1.1 Cost Reduction of PV

Currently, the PV market is one of the fastest growing renewable energy technology markets. The global installed PV capacity has multiplied by a factor of 37.44 in ten years from 1.8 GW in 2000 to 67.4 GW at the end of 2011with a growth rate of 44% per year. In the year 2013, more than 39 GW added. This makes the world wide total capacity to be 139 GW [38]. Despite the rapid growth of the PV market, less than 0.2% of global electricity production is generated by PV. This is because the PV energy costs are typically higher than that from traditional sources such as coal and natural gas power plants [38]. Indeed, the PV systems price has dropped precipitously in recent years and is expected to continue declining in the future, led by substantial reductions in global PV module prices [39, 40]. Unlike traditional energy-production technologies that have ongoing consumables costs, nearly all of the costs for photovoltaic (PV) systems must be paid at the beginning. Reducing those initial capital costs is crucial to reducing the cost of solar electricity. In addition to module price, many factors contribute to the price of a PV system, including installation labor, power electronics, permitting and other regulatory costs, and in the case of ground-mount systems site acquisition and preparation costs [39].

Driven by technological improvements in solar PV modules, manufacturing advances, economies of scale and reductions in balance of system (BoS) costs, the global weighted-average installed costs of utility-scale PV systems could fall by 57% between 2015 and 2025 [41]. Larger cost reductions are possible if deployment accelerates and a more rapid shift to best practice BoS costs occurs. The global average installed cost of utility-scale solar PV could fall by more than half in the next ten years, driven by continued technological improvements, competitive pressures and economies of scale, but also by the convergence of BoS costs towards best practice levels. The majority (about 70%) of the cost reductions will come from lower BoS costs reflecting the high average level of BoS costs today relative to best practice. Module costs are expected to fall by around 42% by 2025, but a narrower cost spread in different markets today means that there are no large gains from cost convergence, unlike for BoS [37].

The biggest cost reduction opportunities for solar PV modules will happen at either end of the crystalline silicon module value chain. Cheaper polysilicon production will halve polysilicon costs per watt by 2025 and account for one-third of the total module cost reduction potential. This will occur along with increased reactor capacity. As a result of the high share of BoS costs today on average, globally, the bulk of the total PV system installed cost reduction potential in the next decade will come from continuous BoS cost reductions. The possible reductions in installed costs could see the LCOE of utility-scale PV projects fall by an average of 59% between 2015–2025 [37].

#### 1.5.1.2 Cost Reduction of CSP

The total installed costs for CSP plants could fall by between 33 and 37% by 2025 [37]. These reductions can be achieved in two ways: by increasing performance and by lowering costs (both capital and operating) as follows [42]:

- Performance can be increased by the following:

  - Improving the solar field optical efficiency.
  - Reducing the solar field thermal losses.
  - Reducing parasitic power consumption.
  - Developing improved configurations that lead to higher utilization and efficiency.
  - Identifying more efficient overall system designs.

- Cost reduction can be achieved by the following:

  - Reducing equipment capital cost via lower material content, lower-cost materials, more efficient design, or less expensive manufacturing and shipping costs.
  - Reducing field assembly and installation costs (via simpler designs and minimization and/or ease of field assembly).

- Lowering operation and maintenance costs via improved reliability, automation, reducing need (as with self-cleaning mirrors), and better techniques.
- Building larger systems that provide economies of scale, particularly in the power block.

By 2025, for a reference CTR plant with nine hours of storage, the total installed costs could decrease from USD 5700/kW in 2015 to USD 3600/kW in 2025 (a 37% reduction) [37]. This will be driven by reductions in the engineering, procurement and construction and owner's cost categories. Together, these two categories will contribute more than half of the expected total cost reduction, while cost reductions in the solar field are expected to account for about one-quarter of the overall reduction potential.

By 2025, the LCOE of CSP technologies could decrease by about 37% for PT plants and by about 44% for CTR. Around 60% of this decrease will be driven by lower installed costs. LCOE reductions out to 2025 will be mostly driven by capital cost reductions, notably for the solar field and thermal energy storage systems, but also by incremental performance improvements. These will come as higher operating temperatures for PT plant, which is unlocked by a transition from synthetic oils as heat transfer fluid to molten salts. This will not only improve the efficiency of the power block, but also will reduce the installed costs of the thermal energy storage system by 40%, as a result of a halving in the required storage volume [37]. CTR plants, with their ability to operate at higher temperatures than parabolic trough systems, have the potential to surpass parabolic troughs as the most competitive CSP option by 2025. By then, PT plants could average USD 0.11/kWh for the reference plant, with CTR achieving costs of USD 0.09/kWh. Lower costs are possible in higher solar irradiation sites than for the reference plant [37].

### *1.5.2 Modeling and Design of CSP Technologies*

As mentioned above, the growing of integrated CSP systems in electrical network creates the need for comprising them in electrical grid reliability studies. There are two techniques for evaluating power system reliability [43]:

- Analytical techniques; more efficient if the operating conditions are not complex.
- Monte Carlo (MC) simulation approach; often preferable when the operating conditions are complex and the number of events is comparatively large.

Therefore, it is worth mentioning that several models and software programs have been proposed to analyze the performance of CSP technologies. Daniel [44] developed an analytical model for 100 kW solar central receiver plant by optimizing the model components such as heliostats size and layout, receiver size, tower height, and turbine engine selection. A complete model for CSP plant without TES system was introduced by Patnode [45], whilst, Price [46] broaden the above-mentioned model by incorporating the TES system to assess the tradeoff between performance, cost, and economic parameters. The performance of solar thermal power plant based on the Brayton cycle with a volumetric air receiver and the Rankine cycle with a tubular water/steam receiver under various locations in Algerian climate was reported by Noureddine et al. [47]. Additionally, Gertig et al. [48] designed a new simulation to estimate the SF of central receiver systems to further simplify the analyses and design of CSP technologies. A mathematical model for a central receiver based on an evacuated solar tube was established to deeply manifest the dynamic characteristics of the developed receive [49]. Zhang and his co-workers [50] investigated Dymola® software to study the dynamic performance of 1 MW central receiver system. In addition, Zhihao et al. [51] adopted an efficient CTR model based on HFLD and TRNSYS software programs to effectively evaluate the CTR performance in China. A simplified configuration of PT solar system and detailed analyses of solar power plant components were indicated by Ricardo [52]. Furthermore, the SAM software was employed to analyze the performance of CSP technologies to simulate the output power of RES [53, 54].

### *1.5.3 ANN Application in Solar Energy Field*

ANN is well known as one of the most commonly used soft computing tool in several different application, i.e., Engineering, Science, Manufacturing, Medicine, etc. [55]. ANN is used to effectively model, simulate, control, optimize and analyze the renewable energy technologies, particularly solar energy systems. Compared to other traditional methods, the ANN soft computing technique provides wide information in multi-dimensional information domains, accurate to solve complex and nonlinear problems, and less time consumed [56]. Presently, ANN technique has been widely used in several applications of renewable energy systems. For example, Boukelia et al. [57, 58] investigated ANN approach for PT solar thermal power plants to predict LCOE of two different integrated PT power plants operated by a thermic oil and a molten salt as primary HTF in each type and probe the techno-economic performances. Harish and Radha applied the ANN method based on different developed models and training algorithms for solar air heater (SAH) aspects such as prediction of exergetic performance of a roughened SAH [59], the thermal performance of unidirectional flow porous bed SAH using different ANN models such as multi-layer perceptron (MLP), generalized regression neural network (GRNN), radial basis function (RBF), and Multiple linear regression [60], heat transfer characteristic obtained from roughened absorber plate to air flowing through SAH ducts [61], and thermal efficiency estimation of wire rib roughened SAH by two different training algorithms [62]. In addition, the thermal performance of solar collectors with flat absorber plate based on ANN model has been declared by Sozen et al. [63]. The optimization of a supercritical organic Rankin cycle coupled with geothermal power plant was further examined by ANN technique with three types of learning algorithms (LM, SCG, and

CGP) [64]. Furthermore, Kalogirou et al. [65–67] utilized ANN approach for sizing PV systems, solar radiation, and wind speed predictions, and performance analysis of solar steam generating plant modeling and design.

## 1.6 Problem Definition

The difficulty design of solar systems, especially central tower receiver technology is when all technical parameters are taken in consideration. Hence, an accurate and simplified computer program was required to get the optimal design parameters that used in LCOE calculation. In addition, the performance of central tower receiver power plant is studied and analyzed using many different models and software programs as reported above, however, these models may be inconvenient to assess the system reliability using MC method due to their raised limitations such as prolonged computational time, complexity, several input parameters, and require more experience and knowledge [68].

Also, the problem of CSP plants is that differ than conventional power plants due to the variable nature of the main energy source (i.e., direct solar radiation), which can't be manipulated. Indeed, the direct beam radiation relies on the weather conditions such as humidity, clouds, and air transparency. Therefore, an effective method is necessary to achieve the required operating temperature despite any variations in the surrounding conditions.

Finally, there is another problem related to the penetration level of PV system that does not include electrical batteries. Therefore, what is the solution to reduce the negative impacts of high penetration of PV system?

## 1.7 Book Objectives

The main objectives of this book include the following proposals:

- Introduce computer programs for optimal design of CTR and PV systems to be interconnected with electric grid. The proposed programs has been designed to determine the optimum number of heliostats field, land area, PV modules number, Inverter units, monthly surplus energy, and monthly deficit energy.
- Using an intelligent system model, using ANN technique, for modeling and simulating CTR power plant in order to predict and analyze its performance and output in a simple and fixable manner.
- Investigating the possibility of using this technique to provide a precise control over the discharge rate of the HTF. Therefore, the receiver outlet temperature remains constant at the desired value regardless of the variations in direct solar radiation and receiver inlet temperature.

- Study the cost analysis of CTR and PV solar technologies to compare their LCOE and find which type has minimum price of generated kilowatt-hour (kWh).
- Studying the integration effect of CSP technology for Aswan site on the penetration level of PV system and system reliability.

## 1.8 Book Outline

To achieve the above objectives, the present book is organized in seven chapters in addition to a list of references. The chapters are organized as follows:

This chapter contains a brief introduction to renewable energy and its advantages, especially solar energy. Also, contains surveying of CSP system technologies and presents a literature review for cost of solar energy, CSP modeling, and the application of ANN technologies in various SF systems. Further, it presents the objectives of the book and the contents of its chapters.

Chapter 2 presents a proposed computer program for optimal design of a CTR system to be interconnected with electric grid. The proposed computer program has been designed to determine an optimum number of heliostats field, land area, and mirror area. The computer program can completely design the CTRPP system interconnected with electric grid and determines the optimum operation hour by hour through the year. Then, it estimates the monthly energy excess and monthly energy shortage. In addition, this study introduces also a proposed program for optimal design of a PV system. The program has been designed to determine an optimum number of PV modules, number of inverters, and land area for the system under study. It estimates the monthly surplus energy and monthly deficit energy. Finally, this study presents an optimal design for CTR and PV hybrid solar systems.

Chapter 3 describes the data and operation method of CTRPP. Also, it presents a simplified mathematical model for all components of CTR plant. This model is used to predict the performance and characteristics of the CTR system even in severe weather conditions. Where, the CTRPP performance depends on many physical parameters like site location, typical weather conditions, the solar radiation incident angle, block factor, cosine factor, and shadow factor. Several models and software programs have been used to analyze the performance of CSP technologies. However, these models may be inappropriate to evaluate the power system reliability. Therefore, this chapter addresses a mathematical modelling for CTRPP from a reasonably simplified model perspective.

Chapter 4 presents the most important models of ANN technique that used in the solar energy fields and the criteria for selecting the optimal model. Also, it offers a new, simple, and accurate method for modeling CTRPP. This technique can control the flow rate of HTF from CST to the tower receiver. Thus, the receiver outlet temperature can be controlled at the required value regardless of the change in solar radiation or the receiver inlet temperature. Additionally, it contains a detailed explanation of the creation steps of the neural network model. Additionally, it presents the results of the proposed model described in Chapter three and four. Comparisons between ANN models to select the optimal model are discussed in this chapter. It was found that the MLP model with LM-40 is the optimal model for controlling the receiver outlet temperature by adjusting the flow rate of the HTF. The proposed model results were compared with the results of SAM simulation program. The results showed full compatibility between them.

Chapter 5 studies the CSP impacts on the penetration level of PV system as well as on the reliability of the system in two cases: The first case is only integration of PV system. The second case is that by using hybrid PV and CSP systems. The results showed that the performance of CSP technologies has a significant positive impact on the system, which supports the overall flexibility of the system because of its ability to store and send generated energy. In addition, the use of CSP reduces the minimum generation constraint of the conventional generators that allows more penetration of the PV system.

Chapter 6 presents the results of the relationship between the energy price generated by the CTR plant with changing the number of storage hours (Ts), solar multiple, and also with the changing capacity of the station. Also, this chapter introduces program for optimal cost and LCOE of CTR system, PV and CTR/PV hybrid solar system. The computer program has been designed to determine optimum design parameters of PV and CTR for the system under study. The decision from the computer program is based on minimum price of the generated kWh from the system. Finally, the objective of this chapter is to decide whether or not a solar PV system is more economical compared to the CTR system. The systems being considered in this study are in Aswan, Egypt as this region has hot and clear weather.

Chapter 7 reports the main conclusions that can be drawn from the book and summarizes the proposed future research topics related to book's work.

## References

1. Owusu PA, Sarkodie SA (2016) A review of renewable energy sources, sustainability issues and climate change mitigation. Cogent Eng 3(1):1167990
2. Shaikh MR, Waghmare SB, Labade SS, Fuke PV, Tekale A (2017) A review paper on electricity generation from solar energy. Int J Res Appl Sci Eng Technol 5(IX):1884–1889
3. Moukhtar I, Elbaset AA, El Dein AZ, Qudaih Y, Mitani Y (2018) Concentrated solar power plants impact on PV penetration level and grid flexibility under Egyptian climate. AIP Conf Proc 1968(1):030037
4. J. Schotte, "Balancing and Storage of Intermittent Renewables: an Economic Study", Master's thesis, University of GENT, 2012.
5. Bejerano JB, Baute ET (2016) Impacts of intermittent renewable generation on electricity system costs. Energy Policy 94:411–420
6. Maioli T (2016) Dynamic analysis and control of a once through steam generator for a concentrated solar power plant. Master's thesis, Polytechnic University of Milan

7. Casati E, Galli A, Colonna P (2013) Thermal energy storage for solar-powered organic Rankine cycle engines. Sol Energy 96:205–219
8. Philibert C (2010) Technology roadmap: concentrating solar power. International Energy Agency's Renewable Energy Division, OECD/IEA, 2010: available at https://www.iea.org/publications/freepublications/publication/csp_roadmap.pdf
9. Denholm P, Mehos M (2011) Enabling greater penetration of solar power via the use of CSP with thermal energy storage. Technical Report NREL/TP-6A20–52978
10. Gielen D (2012) Renewable Energy Technologies: Cost Analysis Series: Concentrating Solar Power. IRENA working paper, Vol. 1, Power Sector Issue (2/5), 2012: available at https://www.irena.org/documentdownloads/publications/re_technologies_cost_analysis-csp.pdf
11. Kalogirou SA (2013) Solar energy engineering: processes and systems. Second Edition, Academic Press
12. Goswami DY (2015) Principals of Solar Engineering, 3rd edn. CRC Press, Taylor & Francis Group
13. Chu Y, Meisen P (2011) Review and comparison of different solar energy technologies. Research Associate Global Energy Network Institute (GENI), San Diego, CA, 2011: available at https://www.geni.org/globalenergy/research/review-and-comparison-of-solar-technologies/Review-and-Comparison-of-Different-Solar-Technologies.pdf
14. Rashad M, El-Samahy A, Daowd M, Amin AM (2015) A comparative study on photovoltaic and concentrated solar thermal power plants. Recent Adv Environ Earth Sci Econ, 167–173
15. Andrews R (2018) Concentrated solar power vs. solar PV, an update. Energy Matters, July 5, 2018: available at https://euanmearns.com/concentrated-solar-power-vs-solar-pv-an-update/
16. Geyer M, Quaschning V (2000) Solar thermal power: The seamless solar link to the conventional power world. Renewable Energy World 3(4):184–191
17. Wolff G, Gallego B, Tisdale R, Hopwood D (2008) CSP concentrates the mind. Renew Energy Focus 9(1):42–47
18. Cardozo FR (2012) Concentrating solar power technologies using molten salts for storage and production of energy. Master Thesis, Faculty of Engineering, University of Porto (FEUP)
19. Dincer I, Midilli A, Kucuk H (2014) Progress in sustainable energy technologies: generating renewable energy. First Edition, Springer
20. Pacheco JE, Wolf T, Muley N (2013) Incorporating supercritical steam turbines into advanced molten-salt power tower plants: Feasibility and performance. SANDIA Report No. 2013–1960
21. Bhandari B, Lee KT, Lee GY et al (2015) Optimization of hybrid renewable energy power systems: A review. Int J Precis Eng Manufact Green Technol 2(1):99–112
22. Chen H, Cong TN, Yang W et al (2009) Progress in electrical energy storage system: a critical review. Prog Nat Sci 19(3):291–312
23. García IL, Álvarez JL, Blanco DJ (2011) Performance model for parabolic trough solar thermal power plants with thermal storage: Comparison to operating plant data. Sol Energy 85(10):2443–2460
24. MENA, "CHAPT ER 1: Review of CSP Technologies", MNA Local Manufacturing Report 4–14–11, 2011: available at: https://www.esmap.org/sites/esmap.org/files/DocumentLibrary/ESMAP_MENA_Local_Manufacturing_Chapter_1.pdf
25. https://solargis.com/maps-and-gis-data/download/middle-east-and-north-africa
26. Behar O, Khellaf A, MohammedI K (2013) A review of studies on central receiver solar thermal power plants. Renew Sustain Energy Rev 23:12–39
27. Mendelsohn M, Lowder T, Canavan B (2012) Utility-scale concentrating solar power and photovoltaics projects: a technology and market overview. Technical Report NREL/TP-6A20–51137
28. Simbolotti G (2013) Concentrating solar power technology brief. IEAETSAP and IRENA Technology Brief E10, 2013: available at: https://www.irena.org/DocumentDownloads/Publications/IRENAETSAP%20Tech%20Brief%20E10%20Concentrating%20Solar%20Power.pdf
29. Sabiha M, Saidur R, Hassani S et al (2015) Energy performance of an evacuated tube solar collector using single walled carbon nanotubes nanofluids. Energy Convers Manage 105:1377–1388

30. Dincer I (2011) Thermal energy storage: systems and applications. 2 edn, John Wiley & Sons, Ltd
31. Herrmann U, Kearney DW (2002) Survey of thermal energy storage for parabolic trough power plants. J SolEnergy Eng 124(2):145–152
32. Gil A et al (2010) State of the art on high temperature thermal energy storage for power generation. Part 1—Concepts, materials and modelization. Renew Sustain Energy Rev 14(1):31–55
33. Seta AL et al (2015) Design of organic Rankine cycle power systems accounting for expander performance. Third International Seminar on ORC Power Systems, Belgium, pp 1–12
34. Saleh B, Koglbauer G, Wendland M, Fischer J (2007) Working fluids for low-temperature organic Rankine cycles. Energy 32(7):1210–1221
35. Macchi E, Astolfi M (2016) Organic rankine cycle (ORC) power systems: technologies and applications. Woodhead Publishing, Elsevier Ltd
36. Murphy C et al (2019) The potential role of concentrating solar power within the context of DOE's 2030 solar cost targets. National Renewable Energy Lab, Golden, Technical Report NREL/TP-6A20–71912
37. Taylor M, Ralon P, Ilas A (2016) The power to change: solar and wind cost reduction potential to 2025. Int Renew Energy Agency
38. EL-Shimy M (2017) Economics of cariable renewable sources for electric power production. LAP LAMBERT Academic Publishing
39. Goodrich A, James T, Woodhouse M (2012) Residential, commercial, and utility-scale photovoltaic (PV) system prices in the United States: current drivers and cost-reduction opportunities. National Renewable Energy Lab.(NREL), Golden, CO (United States)
40. Cook JJ, Ardani K, Margolis R, Fu R (2018) Cost-reduction roadmap for residential Solar photovoltaics (PV), 2017–2030. National Renewable Energy Lab., Technical Report NREL/TP-6A20–70748
41. Osborne M (2016) Balance of system costs key to further solar system cost reductions says IRENA study, available at https://www.pv-tech.org/news/balance-of-system-costs-key-to-further-solar-system-cost-reductions-says-ir
42. Kutscher C et al (2010) Linefocus solar power plant cost reduction plan. National Renewable Energy Lab, Milestone Report NREL/TP-5500–48175
43. Billinton R, Li W (1994) Reliability assessment of electric power systems using monte carlo methods. Springer
44. Murray DJ (2012) Small-scale solar central receiver system design and analysis. Master Thesis, San Luis Obispo University
45. Patnode AM (2006) Simulation and performance evaluation of parabolic trough solar power plants. Master Thesis, University of Wisconsin-Madison
46. Price H (2003) A parabolic trough solar power plant simulation model. International Solar Energy Conference, Hawaii Island, Hawaii, pp 665–673
47. Yamani N, Khellaf A, Mohammedi K, Behar O (2017) Assessment of solar thermal tower technology under Algerian climate. Energy 126:444–460
48. Gertig C, Delgado A, Hidalgo C, Ron R (2014) SoFiA–a novel simulation tool for central receiver systems. Energy Procedia 49:1361–1370
49. Ali BH, Gilani S, Al-Kayiem HH (2016) Mathematical modeling of a developed central receiver based on evacuated solar tubes. MATEC Web of Conf 38:02005
50. Zhang J et al (2014) Dynamic simulation of a 1MWe concentrated solar power tower plant system with Dymola®. Energy Procedia 49:1592–1602
51. Yao Z, Wang Z, Lu Z, Wei X (2009) Modeling and simu lation of the pioneer 1 MW solar thermal central receiver system in China. Renewable Energy 34(11):2437–2446
52. Padilla RV (2011) Simplified methodology for designing parabolic trough solar power plants. PhD Thesis, University of South Florida
53. Aung KT (2011) Simulation tool for renewable energy projects. Technical paper AC2011–1664, American Society for Engineering Education

54. Wagner M, Blair N, Dobos A (2010) A detailed physical trough model for NREL's Solar Advisor Model. Proceedings of the SolarPACES International Symposium, SolarPACES, France, NREL/CP-5500–49368
55. Ghritlahre HK, Prasad RK (2018a) Application of ANN technique to predict the performance of solar collector systems-A review. Renew Sustain Energy Rev 84:75–88
56. Ghritlahre HK, Prasad RK (2017) Prediction of thermal performance of unidirectional flow porous bed solar air heater with optimal training function using artificial neural network. Energy Procedia 109:369–376
57. Boukelia T, Arslan O, Mecibah M (2017) Potential assessment of a parabolic trough solar thermal power plant considering hourly analysis: ANN-based approach. Renew Energy 105:324–333
58. Boukelia T, Arslan O, Mecibah M (2016) ANN-based op imization of a parabolic trough solar thermal power plant. Appl Therm Eng 107:1210–1218
59. Ghritlahre HK, Prasad RK (2018a) Exergetic performance prediction of a roughened solar air heater using artificial neural network. J Mech Eng 64(3)
60. Ghritlahre HK, Prasad RK (2018b) Investigation of thermal performance of unidirectional flow porous bed solar air heater using MLP, GRNN, and RBF models of ANN technique. Thermal Sci Eng Progress 6:226–235
61. Ghritlahre HK, Prasad RK (2018c) Investigation on heat transfer characteristics of roughened solar air heater using ANN technique. Int J Heat Technol 36(1):102–110
62. Ghritlahre HK, Prasad RK (2018d) Development of Optimal ANN Model to Estimate the Thermal Performance of Roughened Solar Air Heater Using Two different Learning Algorithms. Springer, Annals of Data Science, pp 1–15
63. Sözen A, Menlik T, Ünvar S (2008) Determination of efficiency of flat-plate solar collectors using neural network approach. Expert Syst Appl 35(4):1533–1539
64. Arslan O, Yetik O (2009) ANN based optimization of supercritical ORC-Binary geothermal power plant: Simav case study. Applied Thermal Engineering, Vol. 31, No. 17–18, pp. 3922–3928, 2011. A. Mellit, S. A. Kalogirou, L. Hontoria, and S. Shaari, "Artificial intelligence techniques for sizing photovoltaic systems: A review", Renewable and Sustainable Energy Reviews, Vol. 13, No. 2, pp. 406–419
65. Kalogirou SA (2000) Applications of artificial neural-networks for energy systems. Appl Energy 67:17–35
66. Kalogirou SA, Neocleous CC, Schizas CN (1998) Artificial neural networks for modelling the starting-up of a solar steam-generator. Appl Energy 60(2):89–100
67. Gafurov T, Usaola J, Prodanovic M (2014) Modelling of concentrating solar power plant for power system reliability studies. IET Renew Power Gener 9(2):120–130
68. Shen C, He YL, Liu YW, Tao WQ (2008) Modelling and simulation of solar radiation data processing with Simulink. Simul Model Pract Theory 16(7):721–735

# Chapter 2
# Solar Power Plants Design

**Abstract** There are many fundamental differences between CTR technology and other CSP technologies. For example, in the case of PT, solar energy collected by the receiver is directly proportional to the land area of the SF. However, in the case of CTR, solar energy collected by the SF is a complex function of the SF layout respect to the tower. Therefore, for developing a methodology to achieve the optimal design for solar CTR system, it should obtain a method for determining the boundary of the SF around the tower. Hence, this chapter introduces a proposed computer program for optimal design of a CTR system to be interconnected with electric grid. The proposed computer program has been designed to determine an optimum number of heliostats field, land area, and mirror area. The computer program can completely design the CTRPP system interconnected with electric grid and determines the optimum operation hour by hour through the year. Then, it estimates the monthly energy excess and monthly energy shortage. In addition, this chapter introduces also a proposed program for optimal design of a PV system. The program has been designed to determine an optimum number of PV modules, number of inverters, and land area for the system under study. It estimates the monthly surplus energy and monthly deficit energy. Finally, it presents an optimal design for CTR/PV hybrid solar systems.

**Keywords** CTR design · Solar radiation · Solar field boundary · Cosine efficiency · PV design · CTR · PV hybrid system · Generated electrical power · Capacity factor (CF)

## 2.1 Introduction

In the past few decades, In the past few decades, central tower receiver (CTR) technology has been received a great attention as one of the most important types of concentrated solar power (CSP) systems, which is considered a suitable technology for the large solar thermal plants due to its high energy efficiency more than others technologies. There are many fundamental differences between CTR technology and other CSP technologies. For example, in the case of parabolic trough (PT), solar energy collected by the receiver is directly proportional to the land area of the solar field (SF). However, in the case of CTR, solar energy collected by the SF is a complex



I. Moukhtar et al., *Solar Energy*, Power Systems,
https://doi.org/10.1007/978-3-030-61307-5_2

function of the SF layout respect to the tower. Therefore, for developing a methodology in order to achieve the optimal design for solar CTR system, it should obtain a method for determining the boundary of the SF around the tower.

The primary objective of this study is to develop a methodology to carry out a technical analysis of a CTR technology. Therefore, this chapter introduces a proposed computer program for optimal design of a CTR system to be interconnected with electric grid. The proposed computer program has been designed to determine an optimum number of heliostats field, land area, and mirror area. The computer program can completely design the CTR power plant system interconnected with electric grid and determines the optimum operation hour by hour through the year. Then, it estimates the monthly energy excess and monthly energy shortage. In addition, this study introduces also a proposed program for optimal design of photovoltaic (PV) system. The program has been designed to determine an optimum number of PV modules, number of inverters, and land area for the system under study. It estimates the monthly surplus energy and monthly deficit energy. Finally, this study presents an optimal design for CTR and PV hybrid solar systems.

## 2.2 Methodology of CTR Design

In central tower receiver power plant, the first step of its design is the calculation of the solar radiation and sun position considering heliostat and receiver position. The detailed information about solar radiation availability at any location is essential for the design and economic evaluation of CSP solar power plants. Where, lower levels of radiation increase the cost of the CSP plants and therefore their economic viability [1]. Also, the economic feasibility of these plants is based on finding the optimum size for a given electric output.

### 2.2.1 *Solar Angles*

The variation in seasonal solar radiation availability at the surface of the earth can be understood from the geometry of the relative movement of the earth around the sun. The axis of the earth's daily rotation around itself is at an angle of 23.45° to the axis of its ecliptic orbital plane around the sun. This tilt (i.e., declination angle) is the major cause of the seasonal variation of the solar radiation available at any location on the earth [2]. As presented in Fig. 2.1, the sun position can be described at any time by solar altitude angle and solar azimuth angle, which can be calculated as follows [2, 3]:

$$\alpha_s = \sin^{-1}(\sin\delta_s \cdot \sin L_{lat} + \cos\delta_s \cdot \cos L_{lat} \cdot \cos h_s) \tag{2.1}$$

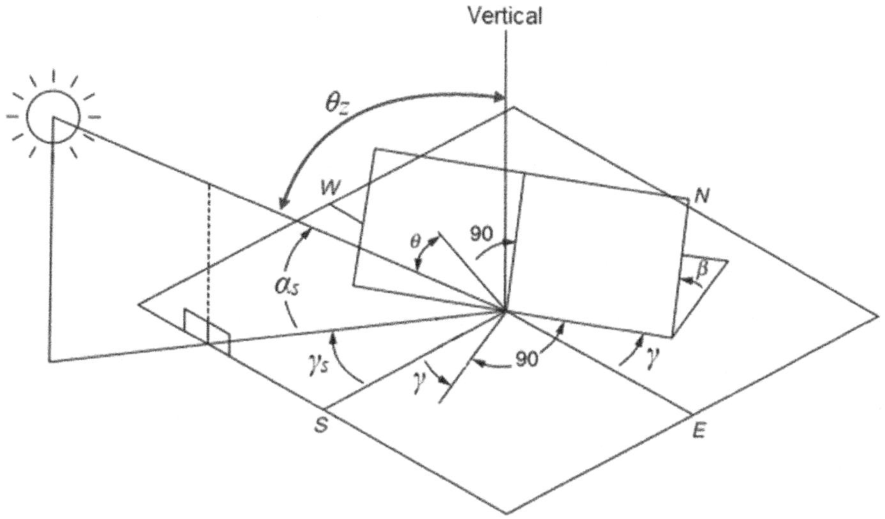


**Fig. 2.1** Definitions of solar angles for a tilted surface [2]

$$\gamma_s = \sin(h_s)\left|\cos^{-1}\left(\frac{\cos\theta_z \sin L_{lat} - \sin\delta_s}{\sin\theta_z \cos L_{lat}}\right)\right| \tag{2.2}$$

where,

$\alpha_s$ solar altitude angle (degree).
$\gamma_s$ solar azimuth angle (degree).
$\delta_s$ solar declination angle (degree).
$L_{lat}$ local latitude (degree).
$h_s$ solar hour angle (degree).
$\theta_z$ solar zenith angle (degree).

The solar altitude angle and azimuth angle must be related to the fundamental angles: solar hour angle, local latitude and solar declination angle. The solar hour angle, is based on the nominal time of 24 h required for the sun to move 360° around the earth or 15° per hour [2], it is defined as:

$$h_s = 15(t_s - 12) \tag{2.3}$$

$$t_s = t + E + 4(L_s - L_{long}) \tag{2.4}$$

where,

$t_s$ solar time (hour).
$t$ local time (hour).

$L_s$ standard time meridian (degree).
$L_{long}$ local longitude (degree).

An approximation for calculating the equation of time is given by [2]:

$$E = 229.2\begin{pmatrix} 0.000075 + 0.001868\cos B - 0.032077\sin B - 0.014615\cos 2B \\ -0.04089\sin 2B \end{pmatrix} \quad (2.5)$$

$B$ is defined as:

$$B = \frac{360}{365}(N - 1) \quad (2.6)$$

where,

$E$ equation of time (minute).
$N$ day number during the year.

The solar declination angle, $\delta s$, is given by [2]:

$$\delta_s = 23.45\left(\frac{360(284 + N)}{365}\right) \quad (2.7)$$

The solar zenith angle is the angle between the site to sun line and the vertical at the site location [2]:

$$\theta_z = 90 - \alpha_s \quad (2.8)$$

The incidence beam angle is an important aspect for the solar energy system design; where the amount of solar thermal energy that could reach the tower receiver depending on this angle. The maximum amount of solar energy at the receiver is decreased by the cosine of this angle [4]. The incidence angle, which depends on the tracking mode and the position of the sun, is calculated in general form for fixed and tracking surfaces as follows [4, 5]:

$$\theta_{inc} = \cos^{-1}(\cos\theta_z\cos\beta + \sin\theta_z\sin\beta\cos(\gamma_s - \gamma)) \quad (2.9)$$

where,

$\theta_{inc}$ incidence beam angle (degree).
$\beta$ surface slope angle (degree).
$\gamma$ surface azimuth angle (degree).

The incidence angle for a plane rotated about a horizontal east–west axis with a single daily adjustment is determined as follows [4]:

$$\theta_{inc} = \cos^{-1}\left(\sin^2\delta_s + \cos^2\delta_s \cos h_s\right) \quad (2.10)$$

In this case, the surface slope angle and surface azimuth angle are given by [4]:

$$\beta = |\mathrm{L_{lat}} - \delta_s|$$

$$\gamma = \begin{cases} 0 & if\ \mathrm{L_{lat}} - \delta_s > 0 \\ 180 & if\ \mathrm{L_{lat}} - \delta_s \leq 0 \end{cases} \quad (2.12)$$

### 2.2.2 Solar Radiation

Solar irradiance is the rate of radiant energy per unit area over a period of time produced from the sun. The units of solar irradiance are W/m$^2$ [4]. Detailed information about solar radiation availability at any location is essential for the design and economic evaluation of central tower receiver power plant. The average amount of solar radiation falling on a surface normal to the rays of the sun outside the atmosphere of the earth, extraterrestrial insolation, at mean earth-sun distance is called the solar constant, $I_o$. Recently, new measurements have found the value of solar constant to be 1366.1 W/m$^2$ [6]. Upon entering the earth's atmosphere, some solar radiation is diffused by air, water molecules, and dust within the atmosphere [4]. The amount of solar radiation reflected, scattered and absorbed depends on the distance travelled by the solar radiation, levels of dust particles and water vapour present in the atmosphere as shown in Fig. 2.2.

DNI represents the portion of solar radiation reaching the surface of the Earth that has not been scattered or absorbed by the atmosphere. Diffuse radiation represents the radiation, which gets scattered by the particles and molecules in the atmosphere, but still made it down to the earth surface, is called diffuse radiation [8]. To convert solar energy into electrical energy by CSP power plants, the minimum threshold DNI necessary for the CSP technology is 2000 kWh/m$^2$/year because of economic constraints [9]. There are many mathematical approaches for estimating the solar radiation. Daily integration (DI) approach is one of the accurate methods that used to calculate the hourly solar radiation [2].

The total radiation on a tilted surface is the sum of the direct beam, diffuse and ground reflected radiation as described in Fig. 2.3. It is obtained from Eq. (2.13) [2].

$$I_{sol} = \left(r_t \overline{H}_h - r_d \overline{H}_d\right) \frac{\cos\theta_{inc}}{\sin\alpha_s} + r_d \overline{H}_d \cos^2\frac{\beta}{2} + \rho_g r_t \overline{H}_h \sin^2\frac{\beta}{2} \quad (2.13)$$

where,

$I_{sol}$ total solar radiation (W/m$^2$).

$\hat{H}_h$ long-term average daily total irradiation on a horizontal surface.

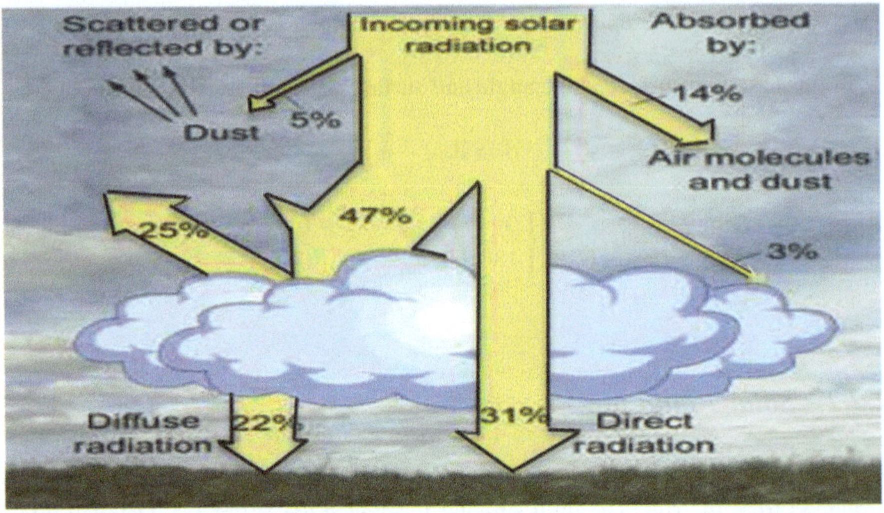


**Fig. 2.2** Attenuation of solar radiation [7]

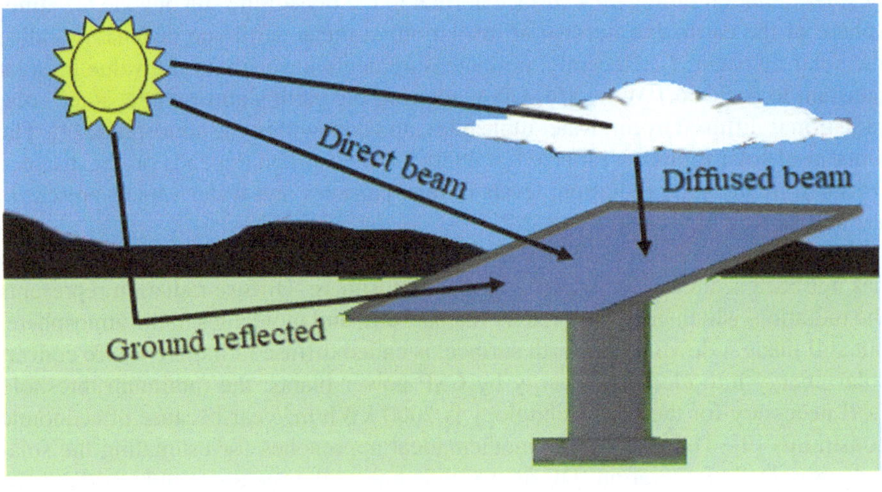


**Fig. 2.3** Components of the global solar radiation [10]

$\hat{H}_d$ long-term average daily diffuse irradiation on a horizontal surface.

$r_t$ ratio of hourly total to the long-term average daily total irradiation on a horizontal surface.

$r_d$ ratio of the hourly diffuse to the long-term average daily diffuse irradiation on a horizontal surface.

$\rho_g$ ground reflectance factor.

$r_t$ and $r_d$ parameters are given in appendix A, while $\hat{H}_h$ and $\hat{H}_d$ are obtained from NASA's Applied Science Program [11].

### 2.2.3 *Solar Field Boundary*

In the case of PT, the SF is uniform everywhere, but in the case of CTR, the optimum SF is closely coupled to the tower height, the latitude of plant location, receiver type (a cylindrical receiver is used in this work, which represents a general pattern), and field layout. Therefore, the determination of the SF boundary with respect to the tower is not straightforward in the case of CTR. There is a method to determine the SF boundary in terms of variation of non-dimensional ratio between the radial distance from the tower ($r$) and tower height ($h_t$) with respect to the azimuth angle. The heliostats efficiencies of CTR system depend on the location of each heliostat relative to the tower receiver as well as the sun's position. From this point, the contours of annual cosine efficiency, annual solar energy per unit mirror area and annual solar energy per unit land area were studied to find SF boundary.

#### 2.2.3.1 Cosine Efficiency Calculation

The most significant loss in the heliostat field is due to the angle between the incident solar beam radiation and a vector normal to the surface of the heliostat (incidence angle) [12]. The effective reflection area of the heliostat is reduced by the cosine of incidence angle, which is called the cosine effect. This may be visualized by considering heliostats at two positions in a field as shown in Fig. 2.4. Heliostat 'H1' situated in the north field has a small cosine loss since its surface normal is almost pointing toward the receiver. Heliostat 'H2' situated in the south field has a higher

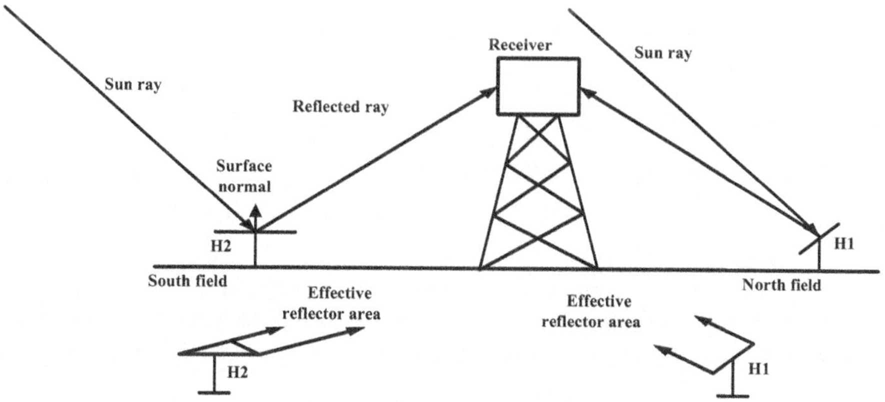


**Fig. 2.4** Cosine factor effect for two heliostats in opposite directions

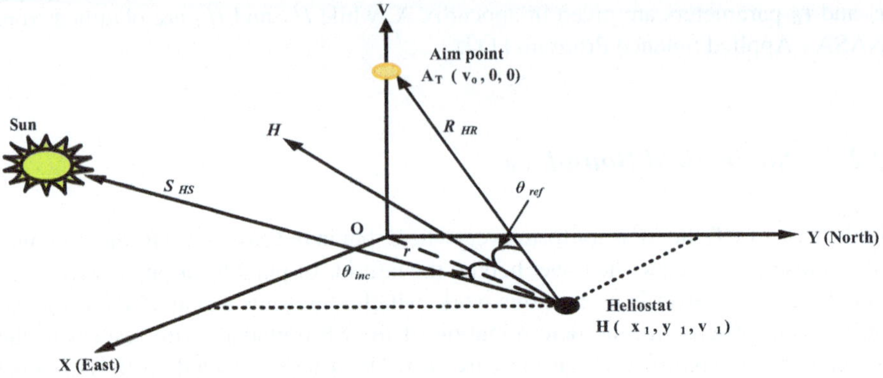


**Fig. 2.5** Sun, heliostat, and tower geometry for calculating $\eta_{cos}$

incidence angle and, consequently, less effective reflector area [4]. Note that the most efficient heliostats are located opposite the sun. An expression for calculation of the cosine of this angle has been developed as Eqs. (2.14) to (2.17) by incorporating the appropriate tower and heliostat position coordinates as defined in Fig. 2.5 [4, 13]:

$$\cos\left(\theta_{inc} + \theta_{ref}\right) = \cos(2\theta_{inc}) = S_{HS} \cdot R_{HR} \tag{2.14}$$

where,

$\theta_{ref}$ reflected beam angle (degree).
$R_{HR}$ heliostat to receiver unit vector.
$S_{HS}$ heliostat to sun unit vector.

$$\cos(2\theta_{inc}) = \frac{(v_0 - v_1)\sin(\alpha_s) - x_1\cos(\alpha_s)\sin(\gamma_s) - y_1\cos(\alpha_s)\cos(\gamma_s)}{\sqrt{(v_0 - v_1)^2 + x_1^2 + y_1^2}} \tag{2.15}$$

$$\cos(2\theta_{inc}) = \frac{\sin(\alpha_s) - \frac{x_1}{(v_0 - v_1)}\cos(\alpha_s)\sin(\gamma_s) - \frac{y_1}{(v_0 - v_1)}\cos(\alpha_s)\cos(\gamma_s)}{\sqrt{1 + \left(\frac{x_1}{v_0 - v_1}\right)^2 + \left(\frac{y_1}{v_0 - v_1}\right)^2}} \tag{2.16}$$

where,

$v_o$ Location of an aim point, A, above the origin point, O, in 'V' direction.
$v_1$ Height of a heliostat mirror, H, above the origin point, O, in 'V' direction.
$x_1$ and y1 Location of a the heliostat mirrors, H, in east and north direction, respectively.

Hence, the cosine factor is calculated according to (2.17) as follows:

$$\eta_{\cos} = \cos\theta_{inc} = \sqrt{\frac{1 + \cos(2\theta_{inc})}{2}} \tag{2.17}$$

For each point of the field fractional annual cosine factor is calculated by [14]:

$$\eta_{\cos,fr} = \frac{\sum_{h=1}^{8760} \cos\theta_{h,p}}{\max\left(\sum_{h=1}^{8760} \cos\theta_{h,p}\right)} \tag{2.18}$$

where,

$\eta_{\cos,fr}$ Fractional annual cosine factor.
$\cos\theta_{h,p}$ Cosine factor at each point of heliostats field for each hour.

Field cosine factor contours have been plotted at three solar altitude angles of 30°, 60°, and 90° as shown in Fig. 2.6.

Figure 2.6 also shows that the heliostats opposite the sun are the most efficient. This is why most of the heliostats in a typical heliostat field will be north of the tower. In the morning, heliostats located in the west direction of the tower will have high efficiency and that located in the east direction of the tower will have low efficiency. The opposite occurs in the afternoon, giving the east and west fields an average efficiency in between the high and the low.

#### 2.2.3.2 Optimum Number of Heliostats Field

For a certain distribution of heliostats in the SF, the total annual reflected solar power per mirror area, at a point $p$ in the SF, is simply as follows [16, 17]:

$$P_m = \sum_{h=1}^{8760} I_{sol,h} \cdot \cos\theta_{h,p} \tag{2.19}$$

where,

$P_m$ Annual reflected solar power per mirror area.

Actually, the heliostats are arranged in a particular layout with gaps between them for minimizing the shadowing and blocking effect, maintenance of mirrors, etc. Therefore, the effect of packing density ($P_d$) will have to be taken into account to see how much area is actually covered by mirrors [4, 14]:

$$P_d = \frac{Mirror\ area}{Land\ area} \tag{2.20}$$

The variation of packing density as a function of ($r/h$) is determined by [15]:

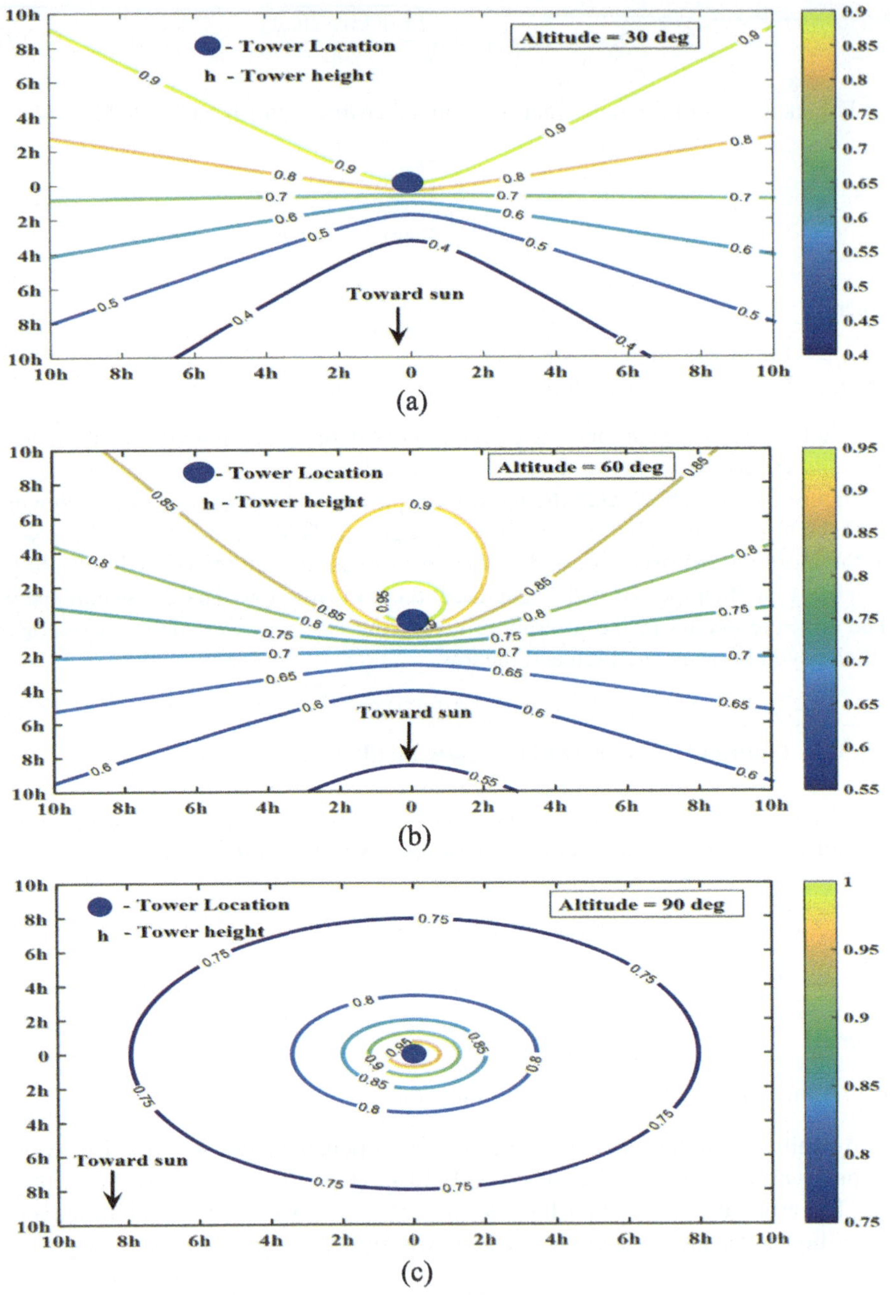


Fig. 2.6 Heliostats cosine factor at: **a** Solar altitude angles of 30°, **b** Solar altitude angles of 60°, and **c** Solar altitude angles of 90°

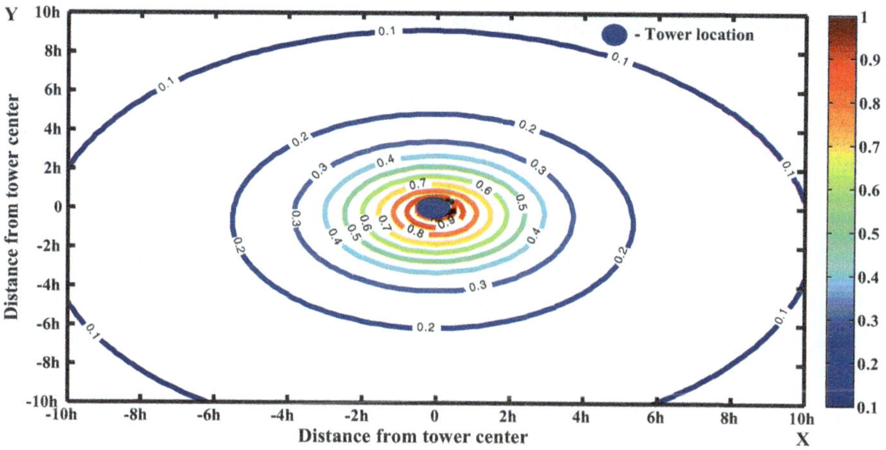


**Fig. 2.7** Contours of energy per unit land area in $MWh/m^2$

$$P_d = \begin{cases} P_d = 0 & if\ \frac{r}{h_t} < \left(\frac{r}{h_t}\right)_{\min} \\ P_d = 0.492 - 0.0939\frac{r}{h_t} & if\ \left(\frac{r}{h_t}\right)_{\min} \le \frac{r}{h_t} \le 2.8 \\ P_d = \frac{0.6}{\sqrt{\left(\frac{r}{h_t}\right)^2 - 1}} & if\ \frac{r}{h_t} > 2.8 \end{cases} \tag{2.21}$$

where,

$P_d$ packing density.
$r$ non-dimensional ratio between the radial distance from the tower.
$h_t$ Tower height.

The value of $(r/h_t)_{\min}$ varies from 0.5 to 1 as observed from existing solar power plants. Consequently, the annual reflected power per unit land area is given by the following equation [14]:

$$P_{LA} = (P_d)_p \sum_{h=1}^{8760} I_{sol,h} \cdot \cos\theta_{h,p} \tag{2.22}$$

where,

$P_{LA}$ Annual reflected power per unit land area.
$I_{sol,h}$ Solar radiation for each hour in the year. Figure 2.7 shows the contours of energy per unit land area; the contours of $P_{lA}$ are circular. Generally, in this methodology, a contour with a value of $P_{LA} = 0.2\ MWh/m^2$ can be chosen as an initial default value (this value can be changed) to determine the outer SF boundary and heliostats number ($N_h$).

### 2.2.4 Tower Height Calculation

In this study, a steam Rankine cycle is used to run the PB. The efficiency of PB depends on many factors such as the inlet pressure and temperature of steam, condenser pressure, PB capacity and mass flow rate of steam, etc. The variation of PB efficiency with the load value can be represented as follows [16]:

$$\eta_{pb} = \begin{cases} \eta_{pb} = 0.18 + 0.011429P_{e,d} - \frac{2.286\times P_{e,d}}{10^4} & if\ 0 \le P_{e,d} \le 15 \\ \eta_{pb} = 0.38 - \frac{6.531P_{e,d}\times(50-P_{e,d})^2}{10^5} & if\ 15 \le P_{e,d} \le 50 \\ \eta_{pb} = 0.38 & if\ P_{e,d} \ge 50 \end{cases} \quad (2.23)$$

where,

$\eta_{pb}$ PB efficiency.
$P_{e,d}$ Design electrical capacity of the plant (MW).

Therefore, the design thermal power of the HTF input to the PB and required receiver solar thermal power that collected by the heliostat field are calculated as the following [16]:

$$P_{th,d} = \frac{P_{e,d}}{\eta_{pb}} \quad (2.24)$$

$$P_{rec,d} = \frac{P_{th,d}}{\eta_{rec}} \quad (2.25)$$

where,

$P_{th,d}$ Design thermal power of the HTF input to the PB (MW).
$P_{rec,d}$ Required receiver solar thermal power that is collected by heliostat field.
$\eta_{rec}$ Receiver efficiency.

The determination of tower height is a function of the receiver thermal power that is collected by the SF. Jebamalai has given a model for tower height with respect to this receiver thermal power [8]. Currently, tower height is calculated as a function of this power for any hour of the year using the following steps [14].

The receiver solar power that is reflected from the heliostats field at any hour is given by [14]:

$$P_{rec,h} = \left(\left(\frac{dx}{h_t}\cdot\frac{dy}{h_t}\right)\eta_{refl}\sum_{h=1}^{8760}\sum_{p}^{N_h} I_{sol,h}\cos\theta_{,h,p}P_{d,p}\eta_{att,p}\right)(h_t)^2 \quad (2.26)$$

where,

$P_{rec,h}$ Receiver solar power that reflected from the heliostats field at any hour.

$\frac{dx}{h_t}$ Dimensionless 'x' coordinate of a point on the field.
$\frac{dy}{h_t}$ Dimensionless 'y' coordinate of a point on the field.
$N_h$ Heliostats number.
$\eta_{refl}$ Attenuation factor at each field point.Reflectivity of the heliostats.
$\eta_{att,p}$ Attenuation factor at each field point.

**Step 1**: Initially, assume that $\eta_{att,ini}$ = one, therefore, the initial tower height is calculated using Eqs. (2.25) and (2.26) as follows:

$$h_{t,ini} = \sqrt{\frac{P_{rec,d}}{max\left(P_{rec,h}\right)}} \tag{2.27}$$

where,

$h_{t,ini}$ Initial tower height.

**Step 2**: Calculate $\eta_{att}$ at each point p, which depends on whether conditions.
For a hazy day model [4].

$$\eta_{att,p} = 0.98707 - 0.2748 S_{n,p} + 0.03394\left(S_{n,p}\right)^2 \tag{2.28}$$

For a clear day model [4].

$$\eta_{att,p} = 0.99326 - 0.1046\left(S_{n,p}\right) + 0.017\left(S_{n,p}\right)^2 - 0.002845\left(S_{n,p}\right)^3 \tag{2.29}$$

where,

$S_{n,p}$ Slant height of the point, p, from the top of the tower in km and is calculated by [14]:

$$S_{n,p} = \left[\frac{\sqrt{(X/h_t)^2 + (Y/h_t)^2 + 1}}{1000}\right] \times h_{t,ini} \tag{2.30}$$

**Step 3**: Calculate $P_{rec,h}$ from Eq. (2.26) at the new value of $\eta_{att}$.

**Step 4**: Compare the maximum value of $P_{rec,h} with P_{rec,d}$ [14].

$$h_t = \begin{cases} h_{t,ini} + dh_t \text{ (repeat step 2) } if\ max\left(P_{rec,h}\right) < P_{rec,d} \\ h_{t,ini} - dh_t \text{ (repeat step 2) } if\ max\left(P_{rec,h}\right) > P_{rec,d} \\ h_{t,f} = h_t \quad \text{ (stop the loop) } if\ max\left(P_{rec,h}\right) = P_{rec,d} \end{cases} \tag{2.31}$$

where,

$h_{t,f}$ Final tower height at solar multiple (SM) = one.
$dh_t$ Tower height increment.

**Step 5**: Determine the tower height at any SM by the following equation [14]:

$$h_{t,SM} = SM \times h_{t,f} \tag{2.32}$$

where,

$h_{t,SM}$ Tower height at any SM value.
SM Solar multiple, which is the ratio between the actual SF size and its size required to supply the turbine at its nameplate capacity with maximum solar radiation.

### *2.2.5 Calculations of Generated Electrical Power and Required Area*

The actual input solar thermal power to the heat exchanger and generated electrical power at any hour are computed as follows [14]:

$$P_{HE,h} = \left(\left(\frac{dx}{h_t}\cdot\frac{dy}{h_t}\right)\eta_{refl}\eta_{rec}\sum_{h=1}^{8760}\sum_{p}^{N_h} I_{sol,h}\cos\theta_{,h,p}P_{d,p}\eta_{att,p}\right)(h_{t,SM})^2 \tag{2.33}$$

$$P_{e,g} = \eta_{pb} \times P_{HE,h} \tag{2.34}$$

$$E_{e,g} = P_{e,g} \times dt \tag{2.35}$$

where,

$P_{HE,h}$ Actual input solar thermal power to the heat exchanger.
$\eta_{rec}$ Receiver efficiency.
$P_{e,g}$ generated electrical power.
$E_{e,g}$ Annual energy production.
$dt$ Time step, which equal one hour.

The CF is the total energy output over a period of time in hours, divided by the product of the period hours and the rated capacity [18].

$$CF = \frac{Annual\ generation\ (MWh)}{Rated\ power(MW) * 8760(h)} \tag{2.36}$$

The total land area and mirror area are given by [14]:

$$A_{land} = \left[ N_h \left( \frac{dx}{h} \cdot \frac{dy}{h} \right) + \pi \left( \left( \frac{r}{h} \right)_{min} \right)^2 \right] (h_{t,SM})^2 \quad (2.37)$$

$$A_{mirror} = \left( \sum_{p}^{N_h} P_{d,p} \right) \left( \frac{dx}{h} \cdot \frac{dy}{h} \right) (h_{t,SM})^2 \quad (2.38)$$

where,

CF Capacity factor.
$A_{land}$ Land area.
$A_{mirror}$ Mirror area.

## 2.3 Methodology of PV Design

### 2.3.1 *Calculation of Average Power for One PV Module*

Normally, the efficiency or electrical power generated and terminal voltage of PV module depends on solar radiation and ambient temperature [18, 19]. The mathematical equation describing the I-V characteristics of a PV solar cells module are given by [18, 20–22]:

$$I(t) = I_{ph}(t) - I_o(t) \left[ exp\left( \frac{q(V(t) + I(t) \times R_s)}{A \times KB \times T(t)} \right) - 1 \right] - \frac{V(t) + I(t) \times R_s}{R_{sh}} \quad (2.39)$$

where,

$I(t)$ The hourly output current, Amp.
$V(t)$ The hourly output voltage, Volt.
$A$ The ideality factor for p–n junction.
$T(t)$ The hourly temperature, Kelvin.
$KB$ The Boltzman's constant in Joules per Kelvin, $1.38 \times 10^{-23}$ J/k.
$R_{sh}$ Internal shunt resistance, ohm.
$R_s$ Series resistance, ohm.
$q$ The charge of the electron in Coulombs, $1.6 \times 10^{-19}$ C.
$I_o(t)$ The hourly reverse saturation current, Amp. This current varies with temperature as follows:

$$I_o(t) = I_{or} \times \left[ \frac{T(t)}{T_r} \right]^3 \times exp\left[ \frac{q \times E_{go}}{KB \times A \times \left( \frac{1}{T_r} - \frac{1}{T(t)} \right)} \right] \quad (2.40)$$

$I_{ph}(t)$: The hourly generated current of solar cells module. This current varies with temperature according to the following equation:

$$I_{ph}(t) = [I_{SC} + K_I(T(t) - 298)] \times \overline{H}_T(t)/100 \tag{2.41}$$

where,

$T_r$ The reference temperature.
$E_{go}$ The band-gap energy of the semiconductor used in solar cells module.
$K_I$ The short circuit current temperature coefficient.
$I_{or}$ The saturation current at $T_r$, Amp.
$\overline{H}_T(t)$ The average hourly radiation on the tilted surface, kW/m$^2$.
$I_{sc}$ PV cell short-circuit current at 25° C and 100 mW/cm$^2$.

The hourly output of the solar cells module can be calculated by the following equation:

$$P_{pv,out}(t) = V(t)\left[I_{ph}(t) - I_o(t)\left(exp\left(\frac{q(V(t) + I(t) \times R_s)}{A \times KB \times T(t)}\right) - 1\right) - \frac{V(t) + I(t) \times R_s}{R_{sh}}\right] \tag{2.42}$$

From the above equations, it is noticed that the output current and power of a PV module are affected by solar insolation and operating cell temperature.

### 2.3.2 *Calculation of Optimum Number of PV Modules*

The energy balance between the load and the output of PV system must be carried out to compute the optimum number of PV modules, $N_{pv}$. The output power from PV system must satisfy the load power demand. The hourly generated power, $P_{pv,out}(t)$, and hourly load power, $P_L(t)$, are compared with each other. If $P_{pv,out}(t)$ is larger than the load power demand then there is an hourly surplus power, but if $P_{pv,out}(t)$ is smaller than the load power demand then there is an hourly deficit power. At any value of $N_{pv}$, if the summation of hourly surplus power equal to the summation of hourly deficit power then this value of $N_{pv}$ represents the optimum number of PV modules. The following equations have been used to get the optimum number of PV modules [23].

$$\text{If} \quad \sum_{t=1}^{t=8760} \left(N_{pv} P_{pv,out}(t) - P_L(t)\right) > 0 \tag{2.43}$$

Then, number of PV module must be decreased by one module and repeating the foregoing process:

$$\text{If} \quad \sum_{t=1}^{t=8760} \left(N_{pv} P_{pv,out}(t) - P_L(t)\right) < 0 \tag{2.44}$$

Then, the number of the PV modules must be increased by one module and repeating the foregoing process.

$$\text{If} \quad \sum_{t=1}^{t=8760} \left(N_{pv} P_{pv,out}(t) - P_L(t)\right) \cong 0 \tag{2.45}$$

Then, $N_{pv}$ is the optimum number of PV modules satisfies the energy balance condition. The value of $N_{pv}$ has been taken as the optimum number of PV modules.

### *2.3.3 Estimation of the Number of Subsystems*

The number of subsystems depends on the inverter rating and efficiency and also on the size of PV system. To determine the number of subsystems, the required following data must be known:

- Rating and efficiency of the inverter unit.
- Solar cell data (module data).
- The optimum number of PV modules that obtained from energy balance process.

Then, the following steps must be carried out [25]:

**Step (1)**: calculation of the subsystem rating current.

$$I_{in} = \frac{S_{in}}{V_{in} \times \eta_{in}} \tag{2.46}$$

where,

$I_{in}$ Subsystem rating current, A
$V_{in}$ The inverter rating voltage, V.
$S_{in}$ The inverter rating power, VA.
$\eta_{in}$ The inverter efficiency.

**Step (2)**: Calculation of the series and parallel modules required for each subsystem.

$$\text{No. of series modules per each string, } N_{series} = V_{in} // V_C \tag{2.47}$$

where,

$V_c$ module voltage at maximum power point, Volt.

$$\text{No. of parallel strings per each subsystem, } N_{parallel} = S_{in}/(P_{nom} \times N_{series}) \quad (2.48)$$

where,

$P_{nom}$ The nominal peak power for solar cells module, W.

$$\text{No. of PV modules per subsystem, } N_3 = N_{series} \times N_{parallel} \quad (2.49)$$

**Step (3)**: Determination the number of subsystems.

$$N_{subsustem} = N_{pvaptimun}/N_3 \quad (2.50)$$

## 2.4 Methodology of CTR/PV Hybrid System Design

The design of CTR/PV hybrid system interconnected to electric grid depends on dividing the load into two parts between CTR and PV. Mathematical equations describing the modeling of CTR and PV systems are given in Sects. (2.2) and (2.3), respectively. The power generated by PV and CTR systems at any time of '*t*' can be expressed by the following equations [23, 24]:-

$$P_{g,total(t)} = \alpha \times P_{pv,out}(t) + (1 - \alpha) \times P_{eg,CTR}(t) \quad (2.51)$$

For the energy balance, the following conditions must be satisfied:

$$\sum_{t=1}^{T_y} P_{g,total(t)} - \sum_{t=1}^{T_y} P_L(t) \cong 0 \quad (2.52)$$

where,

$P_{g,total}(t)$ The total generated power from CTR/PV.
$\alpha$ The system penetration ratio, 0, 0.1, 0.2, 0.3, …….., 1.
$T_y$ The number of hours through year.

The capacity factor of hybrid CTR/PV system '$CF_{hybrid}$' is calculated as follows:

$$CF_{hybrid} = \frac{Annual\ generation\ Summation\ of\ CTR\ and\ PV}{(Rated\ power\ of\ CTR + Rated\ power\ of\ PV) \times 8760} \quad (2.53)$$

## 2.5 Applications and Results

### 2.5.1 CTR Technology Design

A proposed computer program has been designed for calculating optimum design of CTR. The flowchart of this program is shown in Fig. 2.8.

The input data of this program are as follow:

- Site latitude

Aswan site is located on the south of Egypt, its latitude and longitude are 23.97 °N and 32.78 °E, respectively.

- Solar radiation data

For a reliable assessment of CTR plant, the hourly DNI data from measured ground data over several years are required. For Aswan site, a methodology is used to calculate the hourly average DNI data from the available global and diffuse data that has obtained from NASA's Applied Science Program. Figure 2.9 shows the hourly direct beam radiation over the year seasons as a sample data (e.g. for months January, April, July and October) for Aswan site.

- Hourly load demand

It is assumed here that the load demand varies monthly. This means that each month has daily load curve different from other months. Therefore, there are twelve daily load curves through the year. Figure 2.10 shows the load demand for January, April, July and October.

- Design parameters of CTR technology.

In this study, a CTR technology with a cylindrical receiver has been used. Based on the reference data that summarized in Table 2.1, a CTR plant is designed for the selected site, load demand, and other boundary conditions to determine the optimum SF.

The proposed program outputs are the following:

- Optimal number of heliostats field.

The first main of the proposed computer program are the optimal number of heliostats field, based land area, mirror area, and capacity factor. Table 2.2 shows the proposed program output.

The final output of the proposed program after finding the optimal parameters plant design is the energy purchased and sold from the electric grid. The excess and shortage of power from CTR technology for each month are shown in Table 2.3 and Fig. 2.11. Positive and negative power mean that the power send to electric grid and power taken from electric grid, respectively. Figure 2.12 shows the total energy

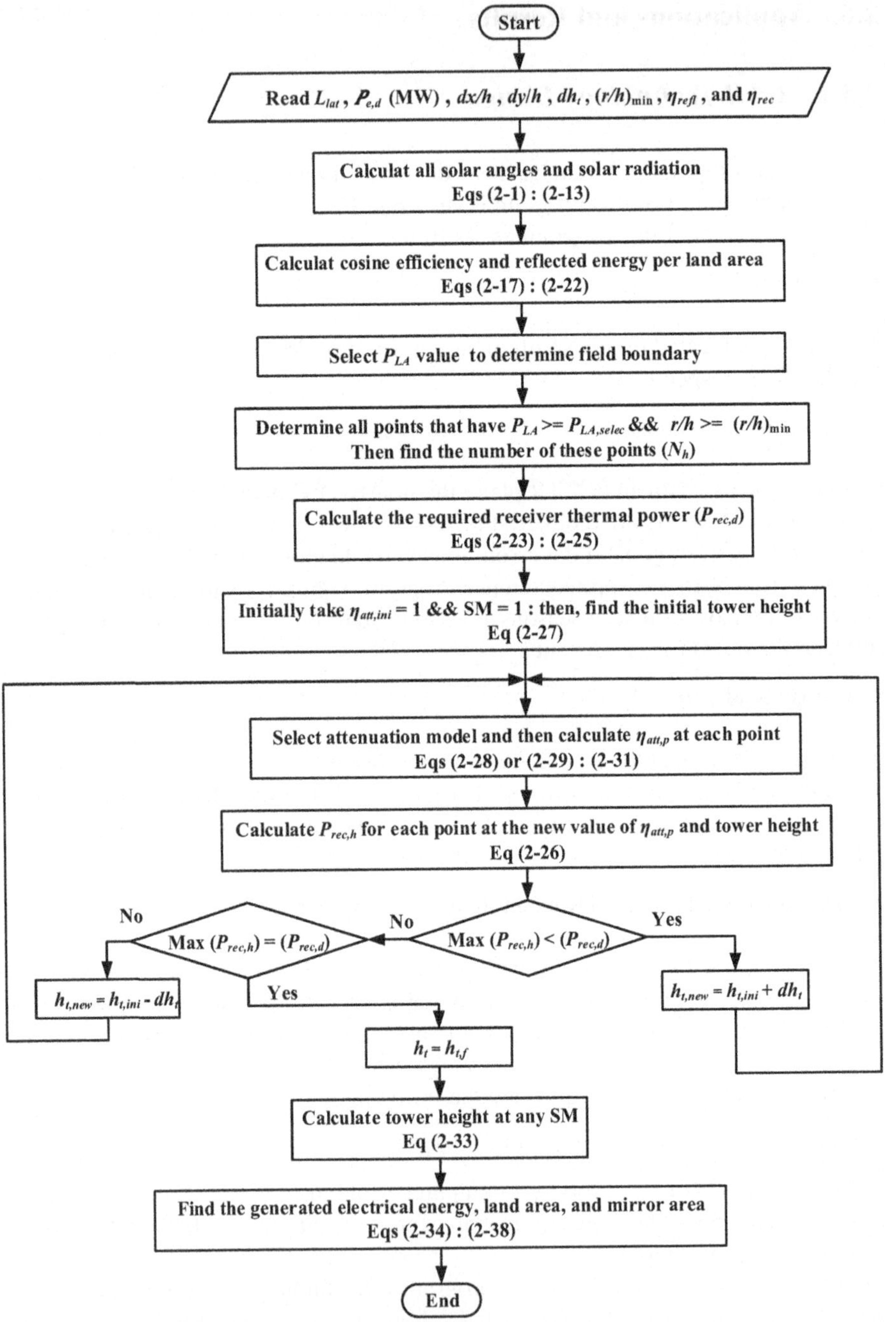


**Fig. 2.8** Proposed computer program flowchart for CTR design

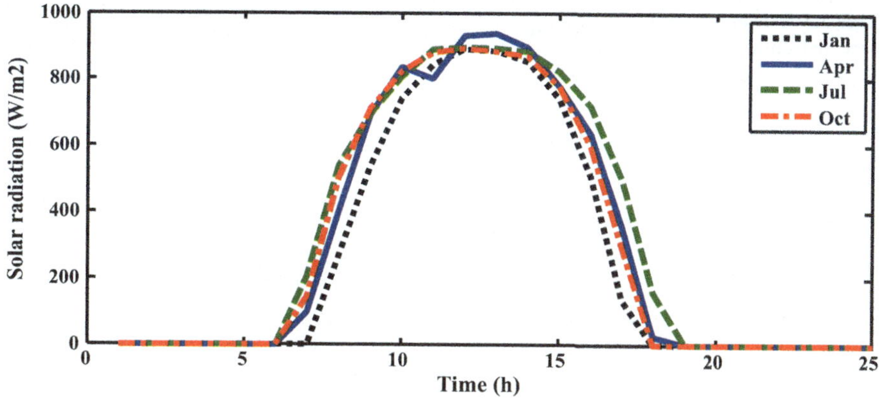


**Fig. 2.9** Solar radiation for Aswan site for Jan, Apr, Jul and Oct

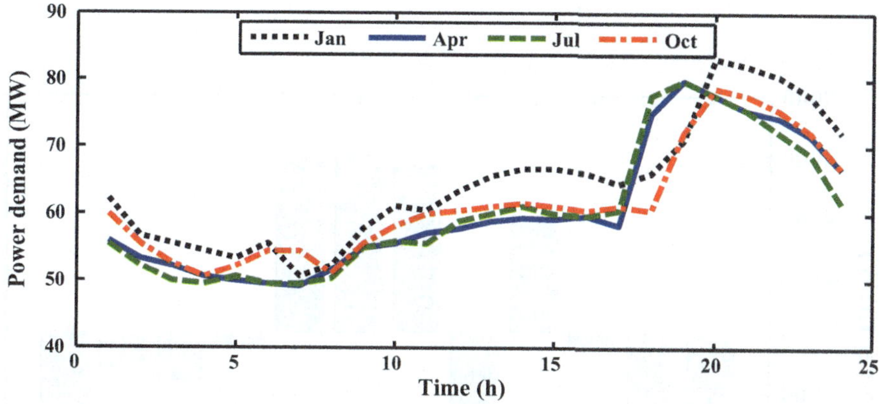


**Fig. 2.10** Daily power demand profile for Jan., Apr., Jul., and Oct.[25]

**Table 2.1** Design data for CTR technology reference system

| Item | $dx/h_t$ | $dy/h_t$ | $(r/h_t)_{\min}$ | $\eta_{refl}$ | $\eta_{rece}$ | $\eta_{st}$ | $W_{hel} * h_{hel}$ ($m^2$) |
|---|---|---|---|---|---|---|---|
| Value | 0.25 | 0.25 | 0.75 | 0.88 | 0.9 | 0.95 | 12.2 × 12.2 |

**Table 2.2** Optimal design of CTR plant

| Plant size (MW) | Solar multiple | Storage hours | Land area (acres) | $A_{mirror}$ ($m^2$) | Heliostats number | CF |
|---|---|---|---|---|---|---|
| 100 | 3 | 9 | 1996 | 1,383,833 | 9585 | 61.63 |

Table 2.3 Power excess and shortage from CTR plant

| Month | Power MW |
|---|---|
| Jan. | – 6,870.34 |
| Feb. | – 3,407.17 |
| Mar. | – 370.75 |
| Apr. | 3,978.98 |
| May. | – 2,707.29 |
| Jun. | 5,728.15 |
| Jul. | 6,206.90 |
| Aug. | 5,921.21 |
| Sept. | – 397.22 |
| Oct. | – 1,625.38 |
| Nov. | – 1,188.64 |
| Dec. | – 5,268.44 |
| Total. | 0.018 |

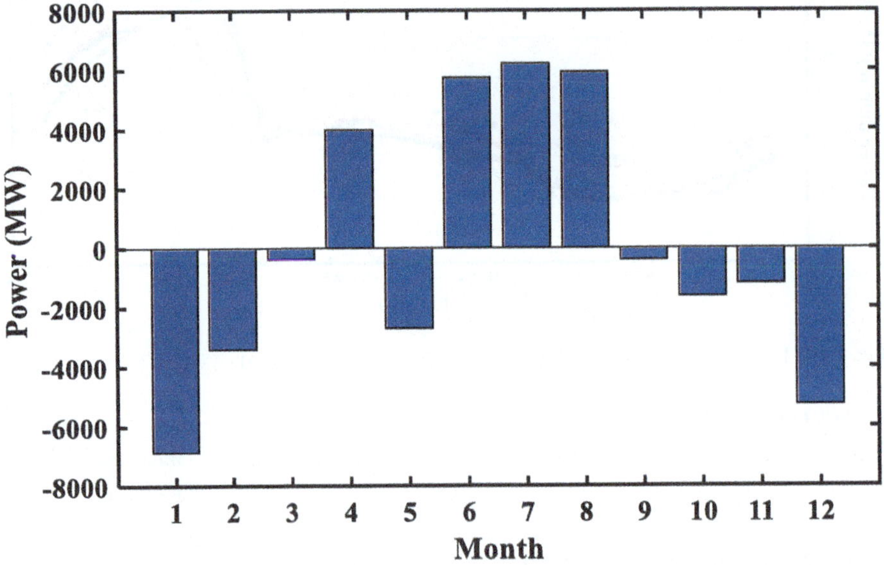


**Fig. 2.11** Total power excess and shortage for each Month

generated form CTR solar plant and energy demand for each month. It can be noticed that the CTR plant with optimal parameters design satisfies the load demand during the year. Therefore, there is a best matching between the load and CTR output.

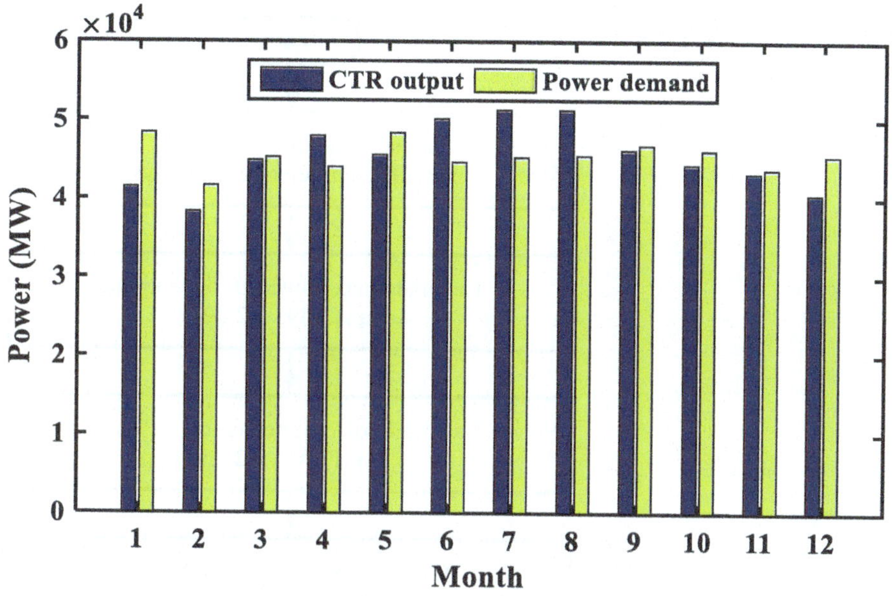


**Fig. 2.12** Total load power and generated power for each Month

## 2.5.2 *PV Technology Design*

A proposed computer program has been designed depending on the above PV design methodology for calculating optimal design of PV system. The flowchart of this program is shown in Fig. 2.13.

**The Program Inputs**:

Table 2.4 shows the monthly average maximum and minimum ambient temperature of Aswan site for months January, April, July and October, which are the input data of this program. Additionally, the characteristics of the used solar cells module is revealed in Table 2.5.

**The Program Outputs**:

This program also gives the optimum total number of PV modules to feed the whole load demand, number of inverter units, area, and capacity factor. Table 2.6 revels the construction of the designed PV system for fixed axis type. In addition, it gives the final output of the PV system which is the energy purchased and sold from electric grid. Figure 2.14 and Table 2.7 show the surplus power and deficit power from PV system for each month. Negative power taken from the grid and positive power send to the grid. Figure 2.15 show the total energy generated form PV system and energy demand for each month. Form Fig. 2.15, it can be seen that the designed PV system satisfies the load demand during the year. Therefore, there is a best matching between the load and PV output.

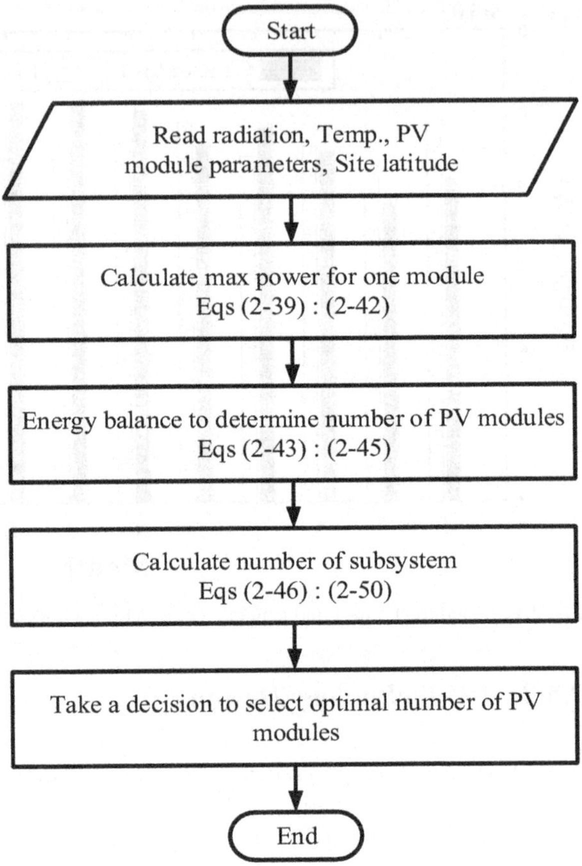


**Fig. 2.13** Flowchart of the proposed program for PV design

**Table 2.4** The maximum and minimum temperature of Aswan site [26]

| Month | Jan | April | July | Oct |
|---|---|---|---|---|
| Max Temperature, C° | 26.2 | 39 | 42 | 41.2 |
| Min Temperature, C° | 7.8 | 15.1 | 25 | 18 |

**Table 2.5** Characteristics of Sunpower SPR-E19-310-COM PV Module [26]

| Area | Nominal Peak Power ($P_{max}$) | Voltage at $P_{max}$, ($V_{mp}$) | Current at $P_{max}$, ($I_{mp}$) | Open circuit voltage at $P_{max}$, ($V_{oc}$) | short circuit current at $P_{max}$, ($I_{sc}$) |
|---|---|---|---|---|---|
| 1.631 m$^2$ | 310.15 W | 54.7 V | 5.7 A | 64.4 V | 6.1 A |

**Table 2.6** Construction of the designed PV system

| Plant capacity (MW) | Capacity factor (%) | Land area (acres) | Module active area, $m^2$ | Total number of modules | Number of inverters units |
|---|---|---|---|---|---|
| 281.167 | 21.99 | 1218 | 1,478,586 | 906,552 | 281 |

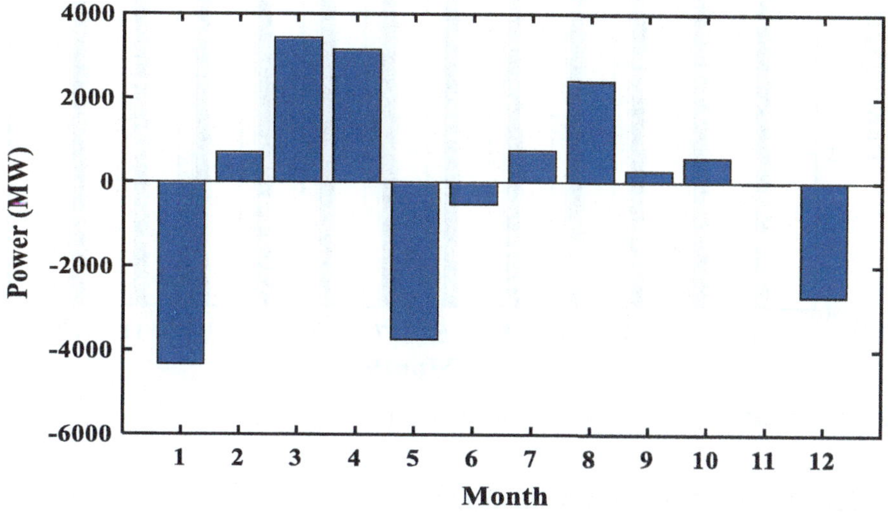


**Fig. 2.14** Total surplus and deficit power for each month

**Table 2.7** Surplus and deficit power from PV System

| Month | Power MW |
|---|---|
| Jan. | – 4,330.30 |
| Feb. | 704.14 |
| Mar. | 3,434.26 |
| Apr. | 3,150.73 |
| May. | – 3,736.25 |
| Jun. | – 521.46 |
| Jul. | 755.30 |
| Aug. | 2,413.14 |
| Sept. | 273.26 |
| Oct. | 583.03 |
| Nov. | – 10.43 |
| Dec. | – 2,715.23 |
| Total | 0.053 |

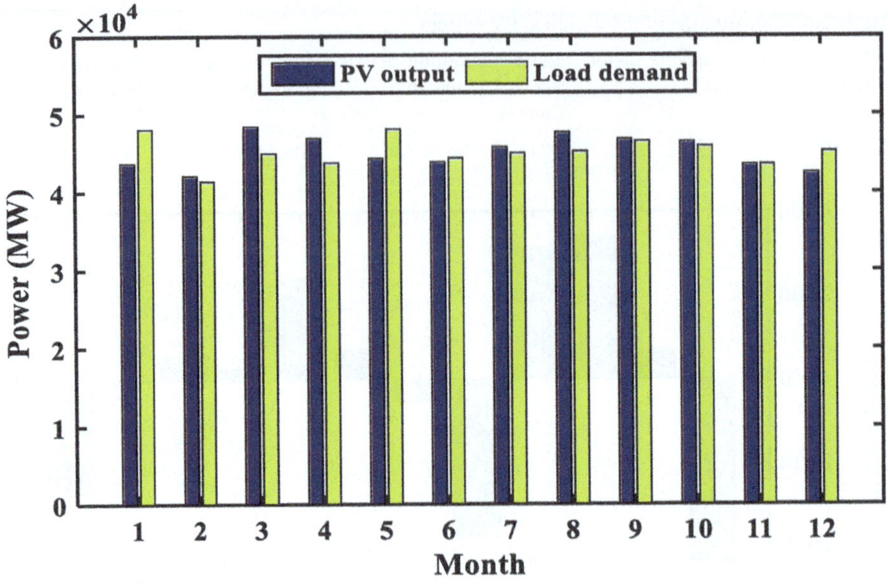


**Fig. 2.15** Total energy generated and energy demand for each month

### *2.5.3 CTR/PV Hybrid System Design*

A computer program has been proposed for design of CTR/PV hybrid system to be interconnected with electric grid. Design of CTR/PV hybrid system can be done by setting penetration ratio in the proposed computer program equal to 0,0.1,0.2,……0.9,1. The penetration level parameters determine the partition of the load. The design algorithm is implemented based on the equations of Sects. 2.2, 2.3, and 2.4 and not only on the energy balance but also depends on dividing the load into two parts between CTR and PV. The results of the proposed program for optimal design of CTR and PV hybrid system have summarized in Table 2.8.

From this table it can be seen that:-

- The lowest used land area and capacity factor occur at penetration level equal to 100% for PV and 0% for CTR.
- The highest used land area and capacity factor occur at penetration level equal to 0% for PV and 100% for CTR.
- For CTR system, the plant capacity of 100 MW with 9 h storage system is required to supply the power demand. While the required plant capacity is 281.16 MW in the case of PV system.
- Finally, the required optimal penetration level not only depends on plant size and required land area, but it mainly depends on the energy price ($/kWh), which will discuss in detail in Chap. 5

**Table 2.8** Impact of penetration level on the optimum design of CTR/PV hybrid system

| Case | Penet. Level, PV (%) | Penet Level, CTR (%) | *PV size (MW)* | *CTR size (MW)* | *PV module no.* | *Heliostat no.* | $A_{land,hyb}$ (acres) | $CF_{hyb}$ |
|---|---|---|---|---|---|---|---|---|
| 1 | 0 | 100 | 0 | 100 | 0 | 9585 | 1996 | 61.63 |
| 2 | 10 | 90 | 28.12 | 90 | 90,648 | 8512 | 1841 | 52.35 |
| 3 | 20 | 80 | 56.23 | 80 | 181,308 | 7475 | 1715 | 45.38 |
| 4 | 30 | 70 | 84.35 | 70 | 271,956 | 6525 | 1669 | 40.07 |
| 5 | 40 | 60 | 112.46 | 60 | 362,616 | 5494 | 1503 | 35.79 |
| 6 | 50 | 50 | 140.58 | 50 | 453,276 | 4581 | 1487 | 32.44 |
| 7 | 60 | 40 | 168.70 | 40 | 543,924 | 3727 | 1530 | 29.68 |
| 8 | 70 | 30 | 196.82 | 30 | 634,584 | 2733 | 1404 | 27.26 |
| 9 | 80 | 20 | 224.93 | 20 | 725,244 | 1805 | 1346 | 25.05 |
| 10 | 90 | 10 | 253.05 | 10 | 815,892 | 936 | 1328 | 23.43 |
| 11 | 100 | 0 | 281.16 | 0 | 906,552 | 0 | 1218 | 21.99 |

# References


1. Padilla RV (2011) Simplified methodology for designing parabolic trough solar power plants. Ph.D. Thesis, University of South Florida
2. Goswami DY (2015) Principals of solar engineering, 3rd edn. CRC Press, Taylor & Francis Group
3. Braun J, Mitchell J (1983) Solar geometry for fixed and tracking surfaces. Sol Energy 31(5):439–444
4. Duffie JA, Beckman WA (2013) Solar engineering of thermal processes, 4th edn. John Wiley & Sons
5. Kalogirou SA (2013) Solar energy engineering: processes and systems, 2nd edn. Academic Press
6. Stine WB, Geyer M (2001) Power from the Sun. https://www.powerfromthesun.net/book.html
7. Gueymard CA (2004) The sun's total and spectral irradiance for solar energy applications and solar radiation models. Sol Energy 76(4):423–453
8. Jebamalai JS (2016) Receiver design methodology for solar tower power plants. Master Thesis, KTH School of Industrial Engineering and Management
9. Breyer C, Knies G (2009) Global energy supply potential of concentrating solar power. Proceedings Solar PACES, Berlin, September, pp 15–18
10. https://agronwww.agron.iastate.edu/courses/Agron541/classes/541/lesson03b/3b.3.html
11. A renewable energy resource web site sponsored by NASA's applied science program. https://eosweb.larc.nasa.gov/sse/
12. Lee JF, Rahim NA, Al-Turki YA (2013) Performance of dual-axis solar tracker versus static solar system by segmented clearness index in Malaysia. Int J Photoenergy 2013:1–13
13. Besarati SM, Goswami DY (2014) A computationally efficient method for the design of the heliostat field for solar power tower plant. Renewable Energy 69:226–232
14. Murray DJ (2012) Small-scale solar central receiver system design and analysis. Master Thesis, San Luis Obispo University
15. Srilakshmi G, Ramaswamy M (2016) Design of solar field and performance estimation of solar tower plants, Center for Study of Science, Technology and Policy (CSTEP), CSTEP-Report-2016–06

16. Collado FJ, Guallar J (2013) A review of optimized design layouts for solar power tower plants with campo code. Renew Sustain Energy Rev 20:142–154
17. Ramaswamy M et al (2012) Engineering economic policy assessment of concentrated solar thermal power technologies for India, Center for Study of Science Technology and Policy, Report CSTEP/E/7
18. Musi R et al (2017) Techno-economic analysis of concentrated solar power plants in terms of levelized cost of electricity. AIP Conf Proc 1850(1):160018
19. Elbaset AA, Ali H, ElSattar MA (2014) Novel seven-parameter model for photovoltaic modules. Sol Energy Mater Sol Cells 130:442–455
20. Elbaset AA, Ali H, ElSattar MA, Khaled M (2016) Implementation of a modified perturb and observe maximum power point tracking algorithm for photovoltaic system using an embedded microcontroller. IET Renew Power Gener 10(4):551–560
21. Elbaset AA, Ali H, ElSattar MA (2015) Modeling of photovoltaic module based on two-diode model, 17th International middle east power systems conference (MEPCON'15). Mansoura University, Egypt
22. Elbaset AA, Ali H, ElSattar MA (2015) Deduction of two-diode model parameters for photovoltaic system, 3rd International Conference on Energy Systems and Technologies, Cairo, Egypt
23. Hua C, Lin J, Shen C (1998) Implementation of a DSP-controlled photovoltaic system with peak power tracking. IEEE Trans Industr Electron 45(1):99–107
24. Elbaset AA (2006) Study of interconnecting issues of photovoltaic/wind hybrid system with electric utility using artificial intelligence. PhD Thesis, Minia University
25. El-Tamaly HH, Elbaset AA (2006) Impact of interconnection photovoltaic/wind system with utility on their reliability using a fuzzy scheme. Renewable Energy 31(15):2475–2491
26. Yousri DM (2011) The Egyptian electricity market: designing a prudent peak load pricing model. Working Paper No. 29, German University in Cairo

# Chapter 3
# Modelling of a Central Tower Receiver Power Plant

**Abstract** To evaluate the performance of a central tower receiver power plant (CTRPP), a model is required to predict its energy output. Normally, the CTRPP performance depends on many physical parameters like site location, typical weather conditions, solar radiation incident angle, block factor, cosine factor, and shadow factor. Several models and software programs have been used to analyze the performance of CSP technologies. However, these models may be inappropriate to evaluate the power system reliability. This chapter describes the data and operation method of CTRPP. In addition, it presents a simplified mathematical model for all components of CTR plant to predict the performance and characteristics of the CTR system. The mathematical modelling for CTRPP is addressed from a reasonably simplified model perspective.

**Keywords** Mathematical modelling · Heliostat field · Receiver solar thermal power · Power block · Steam generator · Operating algorithm scheme

## 3.1 Introduction

To evaluate the performance of central tower receiver (CTR) power plant, a model is required to predict its energy output. So, CTR power plant modeling represents a very important goal in order to predict the performance and characteristics of the CTR system even in severe weather conditions. Normally, the performance of CTR power plant depends on many physical parameters like site location, typical weather conditions, the solar radiation incident angle, block factor, cosine factor, and shadow factor. As mentioned in chapter one, several models and software programs have been used to analyze the performance of CSP technologies. However, these models may be inappropriate to evaluate the power system reliability using Monte Carlo (MC) method. Therefore, this chapter addresses a mathematical modeling for central tower receiver power plant from a reasonably simplified model perspective.



I. Moukhtar et al., *Solar Energy*, Power Systems,
https://doi.org/10.1007/978-3-030-61307-5_3

## 3.2 Description of CTR Power Plant

The capacity of the proposed CTR power plant model is 40 MW under Aswan climate, Egypt (latitude: 23.97 °N and longitude: 32.78 °E). In this chapter, the mathematical model of all CTR's components, which were sketched in Fig. 3.1, is implemented in MATLAB/Simulink as described in Fig. 3.2. The main parameters for CTR modeling are described in Table 3.1. In a molten-salt solar power tower as shown in Fig. 3.1, during daylight periods, the heliostats field collects sun rays and reflects them toward the tower receiver. A liquid salt at 290 °C is pumped from a CST through the receiver where it is heated to 565 °C and then on to a hot tank for storage. When power is needed from the plant, hot liquid salt is pumped to a steam

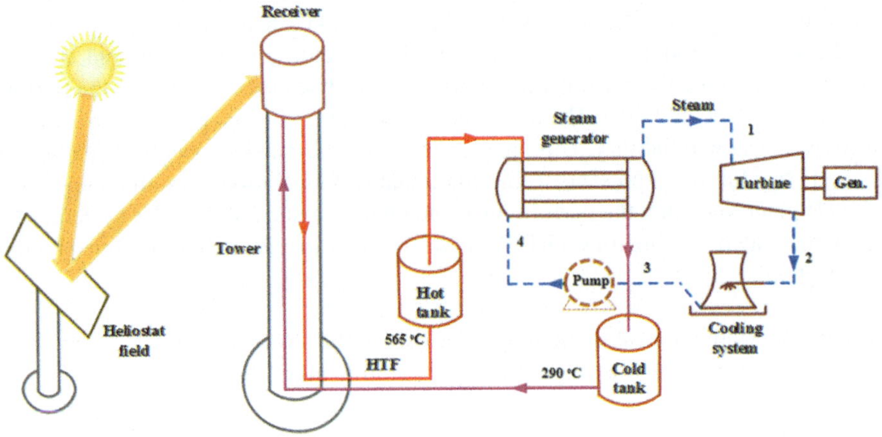


**Fig. 3.1** Schematic diagram of CTR plant components

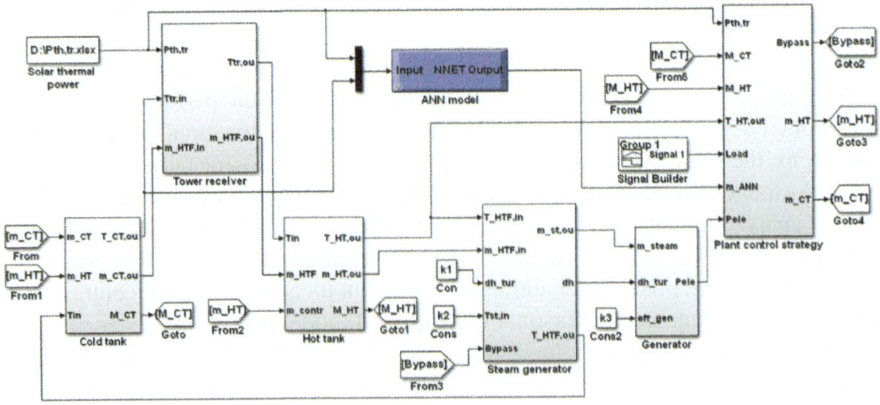


**Fig. 3.2** MATLAB Simulink model of CTR plant

**Table 3.1** Main parameters of CTR model

| Parameters | Value | Parameters | Value |
|---|---|---|---|
| • Local latitude | 23.97 °N | • Generator efficiency | 95 % |
| • Local longitude | 32.78 °E | • Pump efficiency | 80 % |
| • Gross power | 40 MW | • Thermal plant efficiency | 40 % |
| • Inlet HTF temp. to SG | 565 °C | • DNI | 950 W/m$^2$ |
| • Outlet HTF temp. to SG | 290 °C | • Heliostat area | 149 m$^2$ |
| • Turbine inlet steam temp. | 550 °C | • Heliostat reflectivity | 88 % |
| • Turbine inlet pressure | 2 MPa | • Receiver efficiency | 90 % |
| • Turbine outlet pressure | 20 kPa | • Specific heat | 1.6 kJ/kg |
| • Turbine efficiency | 90 % | • Density | 1870 kg/m$^3$ |

generating system that produces superheated steam for a conventional Rankine-cycle turbine/generator system. From the SG, the liquid salt is returned to the CST, where it is stored and eventually reheated in the receiver.

## 3.3 CTR Power Plant Mathematical Modeling

### *3.3.1 Solar Position and Radiation*

This section is the first step for modeling CTR power plant to determine the solar radiation and the sun position considering heliostat and receiver position. The complete calculations and equations of solar angles and solar radiation were discussed in Chap. 2.

### *3.3.2 Heliostat Field*

Heliostats filed or SF represents the main subsystem of CTR power plant and its optical efficiency has a significant impact on the performance of solar power plant. Additionally, as displayed in Fig. 3.3a, b, it represents about 40–50% of the total cost and its annual energy losses are around 40–47% [1, 2]. Consequently, the layout of the heliostats filed is very important step in the modeling and design of CTR power

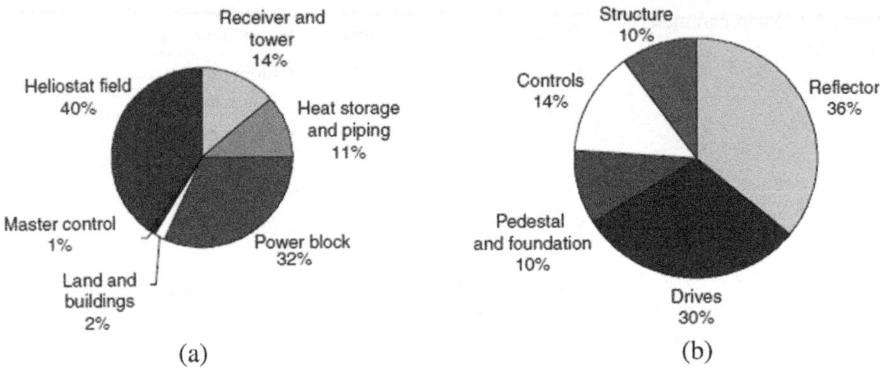


**Fig. 3.3** **a** Investment costs for a CTR plant and **b** Cost of the main components for a single heliostat [4]

plant. In fact, its layout depends on the receiver choice. There are two types of a heliostat field orientation: in first one the surround fields are roughly circular with the heliostats surrounding the tower and in the second one North/south fields have all heliostats on either north or south of the tower [3]. The north field layout (in the Northern Hemisphere) is commonly used for cavity receivers. The north field is more effective at a midday design point and consequently leads to a lower system cost. In this case, the south field performs very well in the morning and evening while it is less effective at midday. On the other hand, the surrounding field is used for cylindrical receivers, which allows the use of a smaller tower [3]. Furthermore, in places close to the equator a surround field would be the best option to make best use of the land and reduce the tower height. North fields improve performance as latitude increases (south fields in the southern hemisphere), in which case, all the heliostats are arranged on the north side of the tower [3].

Due to the heliostat field is the most expensive part and high energy losses in the solar power plant, the heliostats should be carefully distributed in the field to obtain the maximum efficiency. So that any minor improvement in the optical efficiency of the heliostat field will have a positive impact on the overall performance of the system. Indeed, there are several codes have been developed in order to optimize the heliostat field layout for CTR power plant [5, 6]. These computer programs are necessary because both the optimization process of the collector field and the energy evaluation of a given field layout are rather complex problems. System Advisor Model (SAM) is one of the commercial computer programs that can be used for preliminary design and economic analysis, while, there are other programs can be used to design the PB for CTR power plant [3]. Therefore, in this work, SAM program is used for generating heliostat field layout as shown in Fig. 3.4. In CTR systems, the optical performance varies with the heliostat position and therefore its evaluation requires taking into account the layout of heliostat field. Once the relative distribution of each heliostat is known as described in Fig. 3.4, the average optical efficiency can be determined by the following equation [7]:

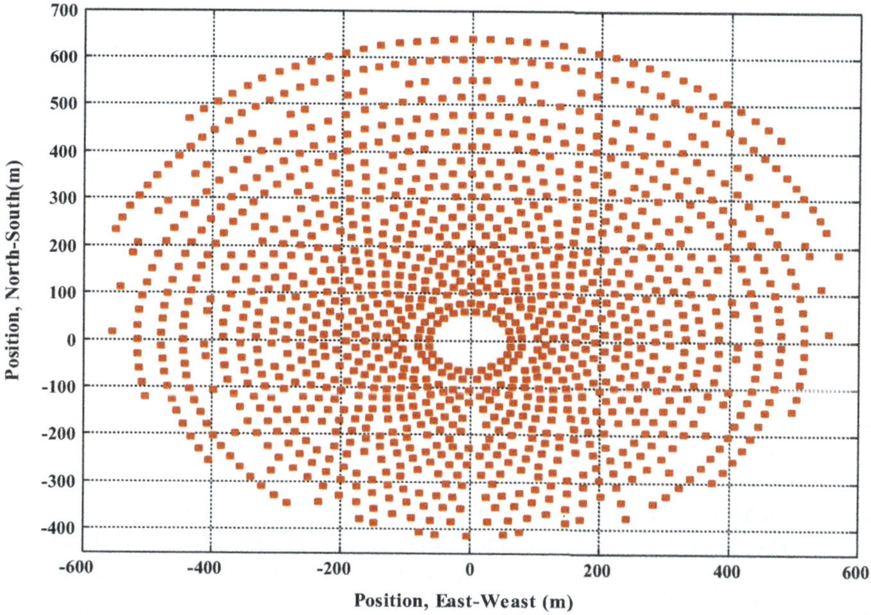


**Fig. 3.4** Heliostat field layout using SAM software

$$\eta_{field} = \eta_{refl} \times \eta_{att} \times \eta_{\cos} \times \eta_{sh\&bl} \tag{3.1}$$

where,

$\eta_{field}$ Field efficiency.
$\eta_{att}$ Attenuation factor.
$\eta_{sh\&bl}$ Shading and blocking factor.

A. **Reflection Factor**

A part of the incident radiation on a heliostat is absorbed by the surface. The fraction that is reflected depends on the absorptivity of the heliostat surface coating and the incidence angle. The value of mirror reflectivity can be set as a constant factor ($\eta_{refl}$ = 0.88) [8].

B. **Cosine Factor and Atmospheric Attenuation Efficiency**

They were discussed in Chap. 2.

C. **Shadowing and Blocking Factor**

The shadowing and blocking factor is defined as the fraction of the area of the heliostat that is free of shadowing and blocking [9]. Shadowing occurs at low sun angles when a heliostat casts its shadow on a heliostat located behind it. Therefore, not

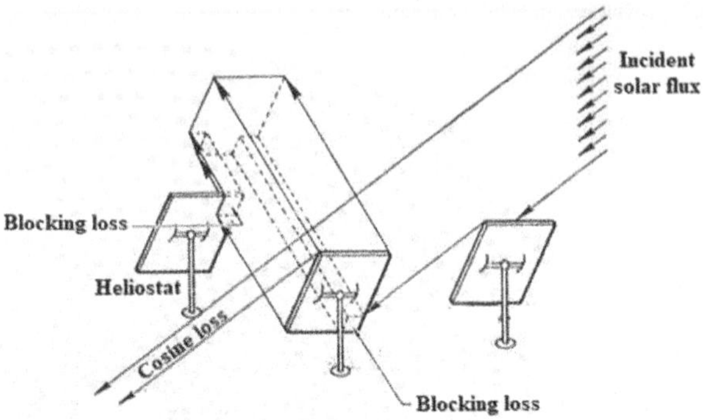


**Fig. 3.5** Shadowing and blocking loss of solar flux [10]

all the incident solar flux is reaching the reflector. Blocking occurs when a heliostat located in front of another heliostat blocks the reflected flux on its way to the receiver [10]. Both processes are illustrated in Fig. 3.5. The efficiency because of shadowing and blocking in heliostat field is of high importance at early and late hours of the day. Even though the shadowing and blocking efficiencies can be maximized by increasing the distance between heliostats, it has to sacrifice the packing factor of heliostat field in which more land area will be needed [11]. In general, its estimation requires complex geometric calculations and detailed description of the field layout. Therefore a proposed simplified equation is used for estimating this efficiency as the following [12]:

$$\eta_{sh\&bl} = 5.52\sin\alpha_s - 12.66\sin^2\alpha_s + 15.13\sin^3\alpha_s - 9.46\sin^4\alpha_s + 2.47\sin^5\alpha_s \tag{3.2}$$

where,

$\alpha_s$ Solar altitude angle.

### 3.3.3 *Receiver Solar Thermal Power*

The solar thermal power reflected by the entire heliostat field into the tower receiver equals to the sum of the product of the field efficiency of each heliostat, heliostat area and the reflected solar radiation by each heliostat.

$$P_{th,tr} = \sum I_{sol} \times A_{hs} \times \eta_{field} \tag{3.3}$$

where,

$P_{th,tr}$ Solar thermal power reflected by heliostats field.
$A_{hs}$ Heliostat area.

The formula which describes the outlet temperature of tower receiver is as follows:

$$T_{tr,out} = \frac{P_{th,tr}\eta_{rec}}{C_{p,HTF}\dot{m}_{CST}} + T_{tr,in} \quad (3.4)$$

where,

| | |
|---|---|
| $T_{tr,out}$ | Tower receiver outlet temperature. |
| $T_{tr,in}$ | Receiver inlet temperature. |
| $\dot{m}_{CST}$ | Outlet mass flow rate from CST. |
| $C_{p,HTF}$ | Specific heat of HTF. |

### *3.3.4 Power Block and Steam Generator Model*

In order to produce electricity, the thermal energy collected by SF must be converted with an appropriate power cycle. Most of thermal power production in the world is based on Rankine cycle and Brayton cycle. Both of them are applicable to solar thermal power conversion, with Rankine cycle being the most popular [3]. Normally, water is the working fluid for the Rankine cycle. However, for lower-temperature solar collection systems (70 °C to approximately 300 °C), an organic fluids are used, in such case the cycle is commonly referred to as ORC [3].

The power cycle begins by collecting the HTF returning from SF in a hot storage system. The HTF is pumped from the hot storage system to a heat exchanger as the energy source for the power cycle. In the heat exchanger, which is normally a shell and tube type heat exchanger, compressed water is heated up until a superheated vapor condition is reached. The superheated steam is expanded through a turbine to operate an electric generator. For simplicity, the detailed design of the heat exchanger (e.g., superheater, SG, and preheater) and turbine stages (e.g., low and high pressure stages) are beyond the scope of this work. However, some features are described here. Initially, mass and energy balance, under steady state conditions, was carried out in each component of the cycle and mass flow rate, temperature and pressure were obtained for each stream.

#### 3.3.4.1 Heat Exchanger

The heat exchanger is used for transferring the heat from the hot HTF to the cold fluid (water) to generate the superheated steam. In this book, the heat exchanger was modeled for partial load conditions in order to accommodate the required electrical power and intermittent energy absorbed by the SF, especially during the periods of cloudy weather and after sunset. These conditions affect the temperature and mass flow rate of the HTF entering the PB. Therefore, the flow rate of outlet superheated steam depends on the required steam power from the heat exchanger. Thus, the HTF inlet mass flow rate is variable to maintain constant steam temperature through the operation cycle. The heat transfer in the heat exchanger is calculated as the following equation [13, 14]:

$$\dot{Q} = UA\Delta T \tag{3.5}$$

where,

$\dot{Q}$ Actual heat transfer rate.
$U$ Overall heat transfer coefficient.
$A$ Heat transfer area respectively.

The effectiveness-NTU method is a better method to calculate the actual heat transfer rate [13, 14].

$$\varepsilon = \frac{\dot{Q}}{Q_{max}} \tag{3.6}$$

where,

$\varepsilon$ Heat transfer effectiveness.
$Q_{max}$ Maximum heat transfer rate between the two fluids.

For a shell and tube heat exchanger and in the case of partial load conditions, the actual heat transfer rate and the heat transfer unit are simplified as in Eqs. (3.7) and (3.8), respectively [15–17].

$$\dot{Q} = \varepsilon Q_{max} = \left(1 - e^{-NTU}\right)\dot{m}_{HST} C_{p,HTF}\left(T_{HTF,in} - T_{steam,in}\right) \tag{3.7}$$

$$NTU = \left(\frac{\dot{m}_{HST,ref}}{\dot{m}_{HST}}\right)^{0.2} NTU_{ref} \tag{3.8}$$

where,

$NTU$ Number of transfer units.
$NTU_{ref}$ Number of transfer units at the reference full-load condition.
$\dot{m}_{HST,ref}$ HTF mass flow rate at the reference full-load condition.
$\dot{m}_{HST}$ Outlet mass flow rate from the HST that equals to inlet mass flow rate to the heat exchanger.
$T_{HTF,in}$ HTF inlet temperature to the heat exchanger.
$T_{steam,in}$ Steam inlet temperature to the heat exchanger.

For further details, refer to Appendix B.

#### 3.3.4.2 Steam Turbine and Generator

In this work, the used power cycle in the model of the adopted CTR plant is a conventional steam Rankine cycle of 40 MW as shown in Fig. 3.1 and the main parameters of the used power cycle are described in Table 3.1. The thermodynamic parameters of Rankine cycle are determined by two main factors [15]: working fluid pressure before the turbine and the pressure in the cooling system. The cooling system pressure is usually determined by the cooling water temperature as the condensation temperature must be higher than the cooling water temperature. The working fluid pressure before the turbine (point 1, as shown in Fig. 3.1) is selected so that the working fluid after the expansion (point 2) is in two-phase condition and its humidity does not exceed a certain value that determined by steam turbine configuration (typically 0.8–0.85). Also, the temperature before turbine must be as high as possible and it is limited by inlet temperature of the HTF. Furthermore, point 3 is usually located on saturation line. Hence, enthalpy after the pump (point 4) is determined by Eq. (3.9) [10].

$$h_4 = h_3 + \frac{P_4 - P_3}{\rho \cdot \eta_{pump}} \tag{3.9}$$

where,

$h_4$ Enthalpy after the pump.
$h_3$ Enthalpy before the pump.
$P_4$ Pressure after the pump.
$P_3$ Pressure before the pump.
$\rho$ Water density.
$\eta_{pump}$ Pump efficiency.

Enthalpy of the point 1 is calculated by according temperature and pressure $h_1 = f(T_1, P_1)$. So, the turbine specific work can be calculated as follows [3]:

$$l_{tur} = \eta_{tur}(h_1 - h_{2s}) \tag{3.10}$$

where,

$h_1$ Actual turbine inlet enthalpy.
$h_{2s}$ Ideal turbine outlet enthalpy.
$l_{tur}$ Turbine specific work (J/kg).
$\eta_{tur}$ Turbine efficiency.
$T_1$ Turbine inlet temperature.
$P_1$ Turbine inlet pressure.

The real process cannot be considered as adiabatic so real enthalpy at the end of expansion process will differ from adiabatic one:

$$h_2 = h_1 - l_{tur} \tag{3.11}$$

where,

$h_2$ Actual turbine outlet enthalpy.

Furthermore, the steam flow rate and the electrical output power of CTR power plant are introduced in Eqs. (3.12) and (3.13).

$$\dot{m}_{steam} = \frac{\dot{Q}}{h_1 - h_2} = \frac{\left(1 - e^{-NTU}\right)\dot{m}_{HST}\left(T_{HTF,in} - T_{steam,in}\right)}{h_1 - h_2} \tag{3.12}$$

$$P_{ele} = \dot{m}_{steam} l_{tur} \eta_{gen} \tag{3.13}$$

where,

$\dot{m}_{steam}$ Steam mass flow rate (kg/s).
$T_{steam,in}$ Turbine inlet steam temperature.
$\eta_{gen}$ Generator efficiency.
$P_{ele}$ Electrical output power of CTR power plant.

#### 3.3.4.3 Thermal Energy Storage

TES is integrated to store heat from SF during daylight and dispatch it during non/low solar radiation time, this strategy increases the value of power generated and number of operation hours or CF compared to stand alone plant. The proposed TES system consists of two tanks of solar salt (60% $NaNO_3$ + 40% $KNO_3$). It's possible to design the storage systems (i.e., CST and HST) by applying the equation of energy balance. The followed criteria to design the storage system are based on the autonomy of the CSP plant, where the goal is to achieve between one and more hours of electrical power production per day without solar radiation. Storage tank size determines the amount of mass within the CST and HST and its capacity (TES full load hours) which refers to the operation hours of the PB provided by thermal storage system. The mass of each tank as a function of the mass flows rate is introduced in Eqs. (3.14) and (3.15).

$$M_{HST}(t) = \int_{t_0}^{t} \dot{m}_{CST} - \dot{m}_{HST}(t)dt + x_0 \tag{3.14}$$

$$M_{CST}(t) = \int_{t_0}^{t} \dot{m}_{HST} - \dot{m}_{CST}(t)dt + x_0 \tag{3.15}$$

where,

$M_{HST}$ Mass inside HST (kg).

$M_{CST}$ Mass inside CST (kg).
$x_0$ Initial condition for the mass (kg) inside the CST and HST.
$t_0$ Lower limit of the two tanks (kg).
$t_u$ Upper limit saturation of the two tanks (kg).

Equation (3.16) is used to determine the stored thermal energy in the TES system [18].

$$E_{stored}(\text{MWh}) = \frac{10^{-6}}{3600} M_{HST} C_{p,HTF} T_{HST,salt}(T_{HST,salt} - T_{CST,salt}) \quad (3.16)$$

where,

$E_{stored}$ Stored thermal energy.
$T_{HST,salt}$ Salt temperature in the HST.
$T_{CST,salt}$ Design salt temperature in the CST.

## 3.4 Operating Algorithm Scheme of CTR Power Plant

The operating strategy of CTR power plant is an important aspect, which ensures the correct operation of each component of this plant under the specified conditions. Figure 3.6 shows a simplified diagram of the plant operation algorithm. The calculation details of this algorithm are given in the previous sections. In this model, a current implemented strategy for deciding plant operation modes depends on solar thermal power, conditions of storage systems, and HST outlet temperature. Most of studies and research in this field carried out a control logic of CTR plant based on the following strategy: firstly, PB is directly fed from SF. Secondly, HST begins discharging for feeding PB when the available SF power is less than the required power [19, 20]. However, in this work, SF power is directly stored in a HST. Then, PB is fed from the HST depending on its conditions (i.e., its temperature and amount of HTF mass within it). This strategy is used to regulate the generated power, which may be useful when a CTR plant is used to feed a certain demand (e.g., supplying loads in desert regions). However, for output power control, there are several advanced optimization and control schemes can be applied to maximize the total benefit and minimize the operating costs [16].

The general aspects of operation modes strategy are summarized as follows:

- Initially, assumed that the state of CST and HST are full and empty, respectively.
- When the solar thermal energy is more than the minimum value of $P_{sol\text{-}min}$ as in Eq. (3.17), the algorithm begins to discharge molten salt from CST that is heated through the receiver and stored in HST.
- Similarly, HST starts to discharge when the mass within HST reaches a predetermined value ($M_{run}$). On the other side, the discharge is stopped when the mass

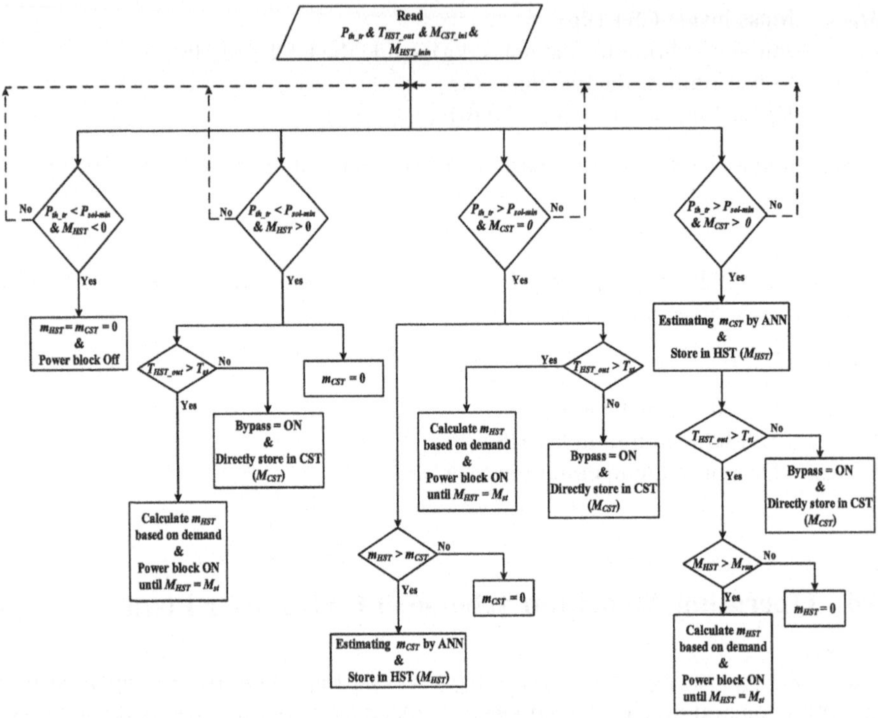


**Fig. 3.6** Schematic diagram of operating modes decisions for CTR/TES plant

reaches a minimum value ($M_{st}$), which makes the pump able to extract any molten salt from the tank.

- Bypass state depends on HST outlet temperature ($T_{HST,out}$). Therefore, the bypass valve opens when $T_{HST,out}$ falls below the temperature setting ($T_{st}$). In this case, the molten salt directly passes from HST to CST and PB does not produce electricity.
- The minimum value of solar thermal energy is given by the following equation [12]:

$$P_{sol-min} = I_{min} A_{hs} \eta_{field} \tag{3.17}$$

where,

$P_{sol\text{-}min}$ Minimum value of solar thermal energy.

# References


1. Behar O, Khellaf A, Mohammedl K (2013) A review of studies on central receiver solar thermal power plants. Renew Sustain Energy Rev 23:12–39
2. Besarati SM, Goswami DY (2014) A computationally efficient method for the design of the heliostat field for solar power tower plant. Renew Energy 69:226–232
3. Goswami DY (2015) Principals of solar engineering, 3rd edn. CRC Press, Taylor & Francis Group
4. Goswami DY, Kreith Y (2015) Energy efficiency and renewable energy hand book, 2nd edn. CRC Press
5. Collado FJ, Guallar J (2012) Campo: generation of regular heliostat fields. Renew Energy 46:49–59
6. Atif M, Al-Sulaiman FA (2015a) Development of a mathematical model for optimizing a heliostat field layout using differential evolution method. Int J Energy Res 39(9):1241–1255
7. Zhou Y, Zhao Y (2014) Heliostat field layout design for solar tower power plant based on GPU vol 47, no 3, pp 4953–4958
8. Wei X et al (2010) A new method for the design of the heliostat field layout for solar tower power plant. Renew Energy 35(9):1970–1975
9. Atif M, Al-Sulaiman FA (2015b) Optimization of heliostat field layout in solar central receiver systems on annual basis using differential evolution algorithm. Energy Convers Manage 95:1–9
10. Stine WB, Geyer M (2001) Power from the Sun. https://www.powerfromthesun.net/book.html
11. Chong KK, Tan M (2012) Comparison study of two different sun tracking methods in optical efficiency of heliostat field. Int J Photo energy 2012
12. Shen C, He YL, Liu YW, Tao WQ (2008) Modelling and simulation of solar radiation data processing with Simulink. Simul Model Pract Theory 16(7):721–735
13. Zohuri B (2017) Compact heat exchangers: selection, application, design and evaluation. Springer
14. Lee HS (2010) Thermal design: heat sinks, thermoelectrics, heat pipes, compact heat exchangers, and solar cells. John Wiley & Sons
15. Patnode AM (2006) Simulation and performance evaluation of parabolic trough solar power plants. Master Thesis, University of Wisconsin-Madison
16. Padilla RV (2011) Simplified methodology for designing parabolic trough solar power plants. PhD Thesis, University of South Florida
17. Kopp JE (2009) Two-tank indirect thermal storage designs for solar parabolic trough power plants. Master Thesis, University of Nevada Las Vegas
18. García L, Álvarez J, Blanco D (2011) Performance model for parabolic trough solar thermal power plants with thermal storage: comparison to operating plant data. Sol Energy 85(10):2443–2460
19. Wagner MJ, Gilman P (2011) Technical manual for the SAM physical trough model. National Renewable Energy Laboratory, Technical Report NREL/TP-5500–51825
20. Wagner MJ (2008) Simulation and predictive performance modeling of utility-scale central receiver system power plants. Master Thesis, University of Wisconsin--Madison

## References

[illegible]

# Chapter 4
# ANN-Based CTR Modelling and Validation Results

**Abstract** Artificial neural network(ANN) is an efficient computing algorithm in MATLAB that emulates the biological neurons performance for the basic functions such as the human brain. Compared to other traditional methods, the ANN soft computing technique provides wide information in multi-dimensional information domains, accurate to solve complex and nonlinear problems, and less time consumed. Presently, ANN technique has been widely used in several applications of renewable energy technologies, particularly solar energy systems. It is used to effectively model, simulate, control, optimize and analyze solar energy systems. Consequently, this chapter presents the most important models of ANN technique that used in the solar energy fields and the criteria for selecting the optimal model. Also, it offers a new, simple, and accurate method for modeling CTRPP. This technique can control the flow rate of HTF from CST to the tower receiver. Thus, the receiver outlet temperature can be controlled at the required value regardless of the change in solar radiation or the receiver inlet temperature. Additionally, it contains a detailed explanation of the creation steps of the neural network model. Additionally, it presents the results of the proposed model described in Chaps. 3 and 4. Comparisons between ANN models to select the optimal model are discussed in this chapter. It was found that the MLP model the optimal model for controlling the receiver outlet temperature by adjusting the flow rate of the HTF. The proposed model results were compared with the results of SAM simulation program. The results showed full compatibility between them.

**Keywords** Artificial neural network(ANN) · Criteria of optimal ANN model · Mass flow rate · Central tower receiver

## 4.1 Introduction

ANN is an efficient computing algorithm in MATLAB that emulates the biological neurons performance for the basic functions such as the human brain. It is a complex information processing system, which is structured with interconnected segmental processing elements called "neurons". These neurons have an ability to determine the nonlinear relationship between the inputs and the outputs, where they perform the following processes [1]: find the input information from other sources, perform


I. Moukhtar et al., *Solar Energy*, Power Systems,
https://doi.org/10.1007/978-3-030-61307-5_4

generally a non-linear operation on the result and then give final results as output. ANN works in two ways, first learning and second storing the knowledge in interconnected links, which is called "weights" [2]. Figure 4.1a describes the basic structure ANN technique, while Fig. 4.1b shows its training process via a comparison between output and target. Consequently, some training processes are performed on ANN in order to become ready for carrying out its function in a self-organized method such as the human brain functions.

The ANN structure consists of three main layers: First, an input layer, which cle onnects the input signal to the neuron via a set of weights. Next, a hidden layer, which summarizes the bias values and the input signals, is weighted by the respective weight values of the neuron. Finally, an output layer is used for limiting the amplitude of the output of the neuron using the activation transfer function. In addition, a bias is added to the neuron to increase or decrease the net output of the neuron [3–5]. The range of ANN outputs depends on the types of the activation transfer function [3, 5].

The major advantages of ANN technique, compared to other computational techniques are its simplicity, high speed and capability to solve complex and nonlinear relationship among the variables and the extracted data. Also, in case when the numerical relations between input and output variables are unknown, and cannot

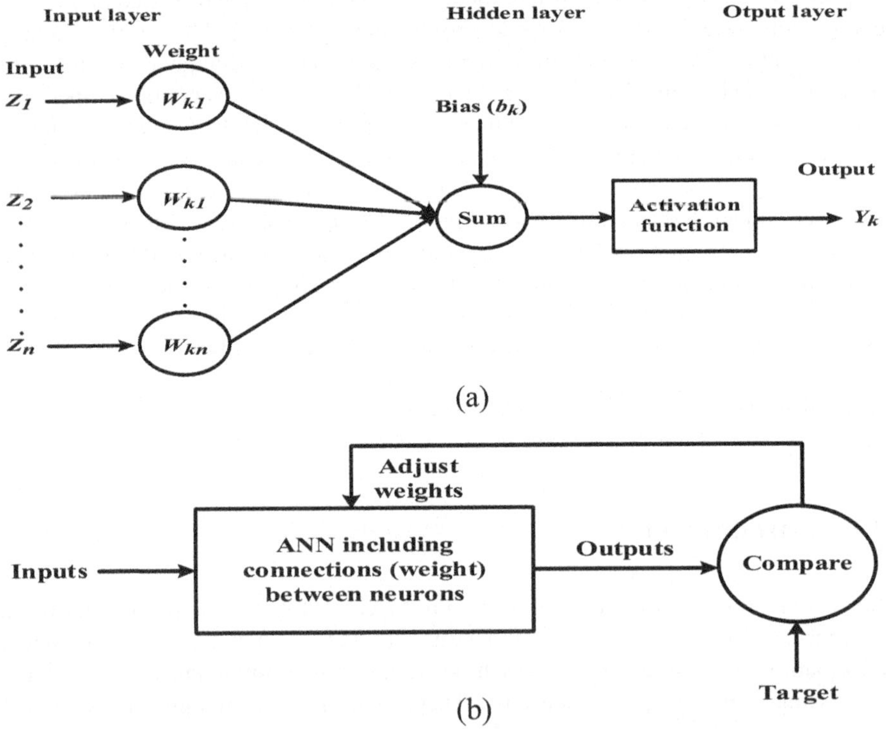


**Fig. 4.1** ANN: **a** Basic structure and **b** Schematic diagram of the basic training process

be incorporated, ANN is found very well suitable for modeling and prediction. The major limitation of the method is, its requirement of the data for training of model, which is not the case with any other analytical methods [2].

## 4.2 ANN Technique Models

In recent years, ANN technique has been becoming increasingly popular in modeling, simulating, controlling, optimizing and analyzing renewable energy technologies, particularly, solar energy systems [2]. Several types or models of ANN are categorised by their structures and abilities. In this work, MLP, RBF, and GRNN were used.

### *4.2.1 MLP Neural Network*

As regard in Fig. 4.2, MLP model consists of three layers: input layer, one or more hidden layers, and an output layer. It is a useful neural network in function approximation [5]. MLP with a single hidden layer can approximate any complex function. Each layer includes a certain number of neurons or nodes, which can determine the nonlinear relationship between the inputs and the outputs [6, 7]. These neurons perform the following processes: receiving the inputs from other sources, combining them, and executing nonlinear operations on the result to give the final output result. Each neuron output is a result of the inputs weighted set.

The sum of bias value and weighted inputs created by neurons are determined as follows [8]:

$$S_0 = \left(\sum\nolimits_{i=1}^{n} W_{ij} Z_i\right) + b_j \tag{4.1}$$

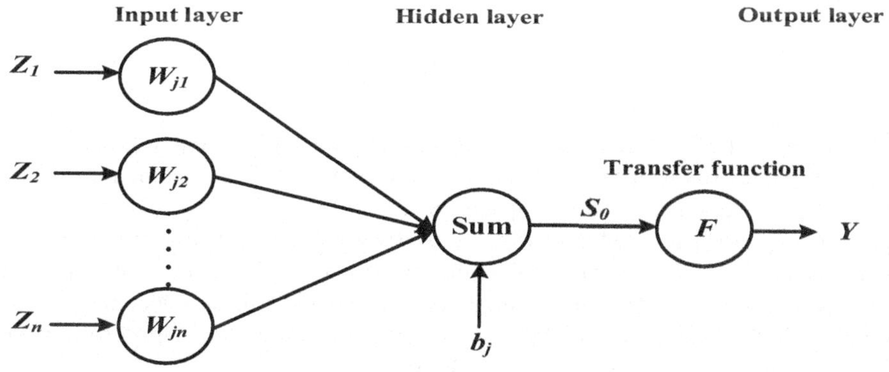


**Fig. 4.2** Basic construction of MLP model

where,

$S_0$: Sum of bias value and weighted inputs created by neurons.
$Z_i$: ANN inputs data.
$W_{ij}$: Neurons weights.
$b_j$: Neuron bias.
$n$: Net input argument.

Then, the sum of $S_0$ passes through a transfer function, which produces an output, is as the following [8]:

$$Y = F(S_0) = F\left[\left(\sum\nolimits_{i=1}^{n} W_{ij} Z_i\right) + b_j\right] \tag{4.2}$$

where,

$Y$: ANN output.
$F$: Transfer function.

The transfer function is consisting of algebraic equation, which is either linear or nonlinear form. Most commonly used transfer functions are log-sigmoid and tangent sigmoid. The log-sigmoid transfer function has been chosen for hidden layer and output layer, if all the values in input and output layer are positive. The inputs and outputs are normalized in 0–1 range. Log-sigmoid transfer is given by [9]:

$$F(z) = \frac{1}{1 + e^{-Z}} \tag{4.3}$$

where,

$z$: Input vectort.

If negative values exist in input or output layer, tangent sigmoid transfer function (tansig) has been preferred. Tangent sigmoid transfer function is given by [9]:

$$F(z) = \frac{e^{Z} - e^{-2Z}}{e^{Z} + e^{-2Z}} \tag{4.4}$$

The network architecture consists of a description of how many layers in the network, number of neurons in each layer, and each layer's transfer function. Therefore, the network architecture impacts on the network training and thence the predicting performance. Moreover, there is no systematic rule to give the optimal neurons number in hidden layer to get the best network performance. In fact, most of researchers have adopted the methodology of trial and error to select the neurons numbers in the hidden layers. On the other hand, several thumb helpful rules were applied to find the neurons number as follows [2]:

$$H_n = \frac{N_I + N_O}{2} + \sqrt{N_T} \tag{4.5}$$

$$H_n = \frac{N_I + N_O}{2} \tag{4.6}$$

$$H_n = \sqrt{N_I \cdot N_O} \tag{4.7}$$

where,

$H_n$: Neurons numbers in hidden layers.
$N_T$: Number of training data.
$N_I$: Number of neurons in input layers.
$N_O$: Number of neurons in output layers.

### 4.2.2 RBF Neural Network

RBF is a functional approximation network, which can be applied in control, memorization and identification like MLP neural network and both of them are feed-forward neural network. It is able to effectively learn system behaviours, therefore, it used for nonlinear systems identification [3]. As demonstrated in Fig. 4.3, RBF is similar to MLP construction, which consists of three layers: input layer, hidden layer, and output layer. In this model, the signal is collected at the input layer and it is passed through the hidden layer. These signals are processed in the hidden layer until they are ready to be sent to the output layer, which generates the output data [10]. The Gaussian transfer function is used to operate the weighted inputs to produce neuron output. The activation function used in hidden layer is radial base function. The RBF for $j$th node in hidden layer is given by Gaussian exponential function as follows [2]:

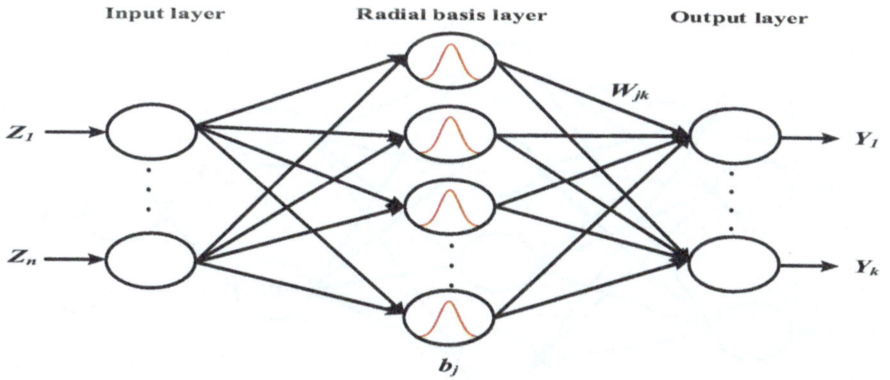


**Fig. 4.3** Basic construction of RBF model

$$b_j(z) = \exp\left[\frac{-\|Z_i - \mu_j\|^2}{2\sigma_j^2}\right] \tag{4.8}$$

where,

$\sigma_j$: Width of $j$th neuron (spread factor).
$j$: Number of neuron in hidden layer.
$\mu_j$: RBF centre unit.

The network output '$Y_k$' for output layer is linear, which is given by [2]:

$$Y_k(z) = \sum_{j=1}^{n} W_{kj} b_j(z) + b_k \tag{4.9}$$

### 4.2.3 GRNN Neural Network

GRNN is also often utilized as function approximation and it is based on a probabilistic model. GRNN includes a radial basis layer and a special linear layer. Subsequently, it is similar to RBF network in its construction but has a slightly different second layer [5]. GRNN is a four-layer feed forward neural network based on the nonlinear regression theory consisting of input layer, pattern layer, summation layer and output layer [11–13]. The configuration of GRNN is illustrated in Fig. 4.4. GRNN is a memory-based feed forward networks based on the estimation of probability density functions. There are no training parameters such as learning rate and momentum factor, as there are in back propagation networks, but there is a smoothing factor applied after the network is trained [9]. The summation layer has two different types of processing units (the summation unit and a single division unit). Each of the GRNN

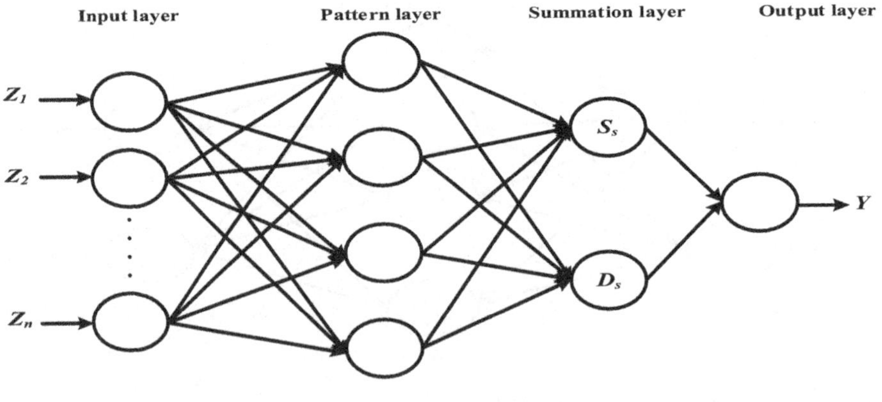


**Fig. 4.4** Basic construction of GRNN model

output units is connected only to its corresponding summation unit and to the division unit. The summation and output layers together basically perform a normalization of the output vector. Radial base and linear transfer functions are used in hidden layer and output layer, respectively.

The information is collected by the input layer and transmitted to the pattern layer that performs clustering on the training process [14]. Then it passes through the summation layer, which includes only two neurons [15]:

$$S_s = \sum_{i=1}^{n} W_i exp[-G(z, z_i)] \tag{4.10}$$

$$D_s = \sum_{i=1}^{n} exp[-G(z, z_i)] \tag{4.11}$$

where,

$S_s$: S-summation neuron in summation layer.
$D_s$: D-Summation neuron in summation layer.
$G(z, z_i)$: Gaussian function and is defined as [15]:

$$G(z, z_i) = \sum_{j=1}^{n} \left( \frac{z_j - z_{ij}}{\sigma_j} \right)^2 \tag{4.12}$$

where,

$z_j$: Represents the $j$th element of $z$.
$z_{ij}$: Represents the $j$th element of $z_i$.

The fourth layer (output layer) executes the output normalization as given in Eq. (4.13) [15].

$$Y(z) = \frac{\sum_{i=1}^{n} W_i \exp[-G(z, z_i)]}{\sum_{i=1}^{n} \exp[-G(z, z_i)]} \tag{4.13}$$

## 4.3 Criteria of Optimal ANN Model

During the training process, the training algorithms minimize the error between the target and the predicted values of the adopted ANN model. The performance index of each training algorithm can be determined by the mean square error (MSE) as the following [16]:

$$MSE = \frac{1}{N_S} \sum_{i=1}^{N_S} \left( Z_{actual,i} - Z_{predicted,i} \right)^2 \tag{4.14}$$

where,

MSE: Mean square error.
$Z_{actual}$: Actual value of the target.
$Z_{predicted}$: Predicted value by the ANN model.
$N_S$: Total number of samples.

Furthermore, the performance of each ANN model was evaluated by numerous statistical parameters such as the coefficient of determination ($R^2$), root mean square error (RMSE), coefficient of variation (COV), and mean absolute error (MAE). The optimal ANN model is that achieves the lowest error (preferred to be 0 or close to 0) and gives the highest value of $R^2$ (expected to reach one or close to one), which are given as [6]:

$$RMSE = \sqrt{\frac{1}{2}\sum_{i=1}^{N_S}\left(Z_{actual,i} - Z_{predicted,i}\right)^2} \quad (4.14)$$

$$MAE = \frac{1}{N_S}\sum_{i=1}^{N_S}\left|Z_{actual,i} - Z_{predicted,i}\right| \quad (4.15)$$

$$R^2 = 1 - \frac{\sum_{i=1}^{N_S}\left(Z_{actual,i} - Z_{predicted,i}\right)^2}{\sum_{i=1}^{N_S} Z_{predicted,i}^2} \quad (4.16)$$

$$COV = 100 \cdot RMSE/\left(\frac{1}{N_S}\sum_{i=1}^{N_S} Z_{predicted,i}\right) \quad (4.17)$$

where,

$R^2$: Coefficient of determination.
RMSE: Root mean square error.
COV: Coefficient of variation.
MAE: Mean absolute error.

## 4.4 ANN Method for Estimating the Mass Flow Rate

### *4.4.1 Structure of ANN Model and Parameters*

In this work, three different models of ANN approach; RBF, GRNN, and MLP, were used for modeling and predicting the output power of the CTR plant. The proposed models were applied to control the HTF mass flow rate from a CST, which passes through a tower receiver. ANN models were implemented with two input parameters, one output parameter, and one hidden layer. The inputs comprise receiver inlet temperature and receiver thermal power. The output includes mass flow rate of the HTF. Therefore, to estimate HTF mass flow rate ($\dot{m}_{CST}$) from the CST to the

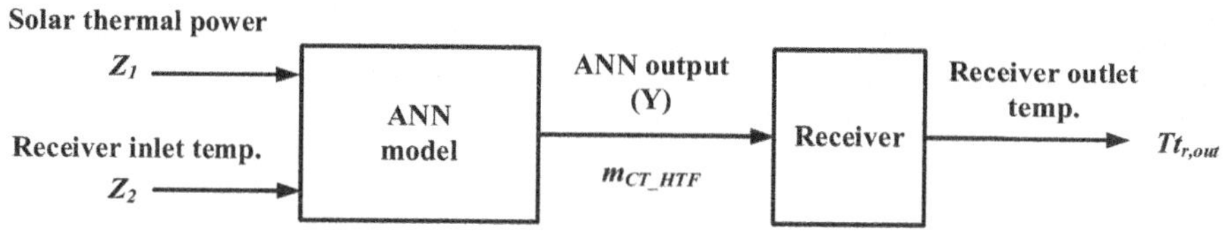


**Fig. 4.5** Block diagram of mass flow rate and temperature control using ANN

tower receiver, solar thermal power ($P_{th,tr}$) and receiver inlet temperature ($T_{in,tr}$) have been considered as the important criteria for the secure solar plant operation. As exemplified in Fig. 4.5, the ANN can control the amount of discharge rate through the two inputs; $Z_1$ and $Z_2$, which represent $P_{th,tr}$ at the tower receiver with a range of 0–150 $MW_{th}$ and $T_{in,tr}$ of the HTF mass flow rate with a range of 290–450 °C, respectively. Where, the HTF mass flow rate ($Y$) is the ANN model output with a range of 0–870 kg/s.

The HTF discharge rate from the CST was calculated by applying the energy balance equation at the tower receiver [17].

$$\dot{m}_{CST} = \frac{\eta_{rec} \times P_{th,tr}}{C_{HTF}\left(T_{tr,in} - T_{tr,out}\right)} \tag{4.18}$$

where,

$T_{tr,in}$: Inlet receiver temperature.
$T_{tr,out}$: Outlet receiver temperature.

Based on the historical calculated data, the proposed ANN model can evaluate the optimal discharge rate in order to stabilize the outlet receiver temperature at desired point during the sudden changes of the weather and/or plant conditions. A total of 888 datasets were obtained by calculating the HTF discharge rate at different points of the receiver inlet temperature. Therefore, 70% of total 888 data sets was taken for training process, while the rest of data was utilized for testing and validation process. Additionally, three learning algorithms; LM, BFG, and SCG, are applied for training the MLP model to determine the best model for estimating the HTF discharge rate. Furthermore, the neurons number was multiplied based on trial and error method from 5 to 40 in the hidden layer in the training process in order to precisely predict the output result. Table 4.1 summarizes the main required parameter for training and testing MLP model. The three models of ANN approach were trained based on Eq. (4.18) in order to estimate the HTF discharge rate. The performance of ANN models were evaluated with a goal equals to $10^{-5}$. The amount of HTF discharge rate is adjusted according to the values of both $Z_1$ and $Z_2$. Consequently, the ANN model regulates the receiver outlet temperature at the designed value, which depends mainly on the following parameters: amount of mass flow rate through the receiver, receiver inlet temperature, and solar intensity concentrated on the receiver.

**Table 4.1** Main parameters of the proposed MLPNN model

| Parameters | Value |
|---|---|
| Input | 2 |
| Output | 1 |
| Hidden layer | 1 |
| Hidden layer transfer function | Logsig |
| Output layer transfer function | Purelin |
| Data sets | 888 for each input |
| Training data | 70% of data sets |
| Testing and validation data | 30% of data sets |
| Goal or error tolerance | $10^{-5}$ |
| Max number of iterations | 1000 |
| Neurons range for training process | 5–40 |

### 4.4.2 Data Preparation

Owing to the difference of parameters units between the input and output, the data rescaling (normalization) is a vital process before developing the ANN models to improve the training convergence of an ANN technique. Basically, the normalization allows the neural network to converge faster and attain better results. Variable normalization is a common practice in ANN modeling, especially when the range of variation is very wide and different. Therefore, normalization (scaling) of data within a uniform range is essential to avoid data with larger magnitude from overriding the smaller ones [18]. Generally, different approaches are used (i.e., min rule, max rule, etc.) for normalizing the variables between 0 to 1 and − 1 to 1 [2]. The normalized value between − 1 and 1 for each raw input/output data set was calculated as [2]:

$$T_{normal} = \frac{T_i - T_{min}}{T_{max} - T_{min}}(High_{value} - Low_{value}) + Low_{value} \tag{4.19}$$

where,

$T_{normal}$: The normalized value of the variable $T_i$.
$High_{value}$: The high value that equal +1.
$Low_{value}$: The low value that equal −1.
$T_i$: Variable which represents the target.
$T_{max}$: Represents the value of the maximum of variable $T_i$.
$T_{min}$: Represents the value of the minimum of variable $T_i$.

While, the normalized value between 0 and 1 for each raw input/output data set was calculated as [19]:

$$T_{normal} = \frac{T_i - T_{min}}{T_{max} - T_{min}} \tag{4.20}$$

### 4.4.3 ANN Model Creation

ANN is trained with set of known input–output data and suitable learning method to perform a function by adjusting the values of weight coefficient between processing neurons. The training process continues until the network output matches with desired output. Changing the weights and biases reduces the error between the network output and desired output. The training process is terminated automatically when the error falls below a determined value. For further clarification as demonstrated in Fig. 4.6, the steps for developing or training the ANN model for estimating the discharge rate is demonstrated as follows [20]:

- Set input and output data
- Divide the data sets for training, testing and validation processes. About 70% of the input–output data selected randomly are assigned as training set data and remaining data can be used for testing and validating the network.

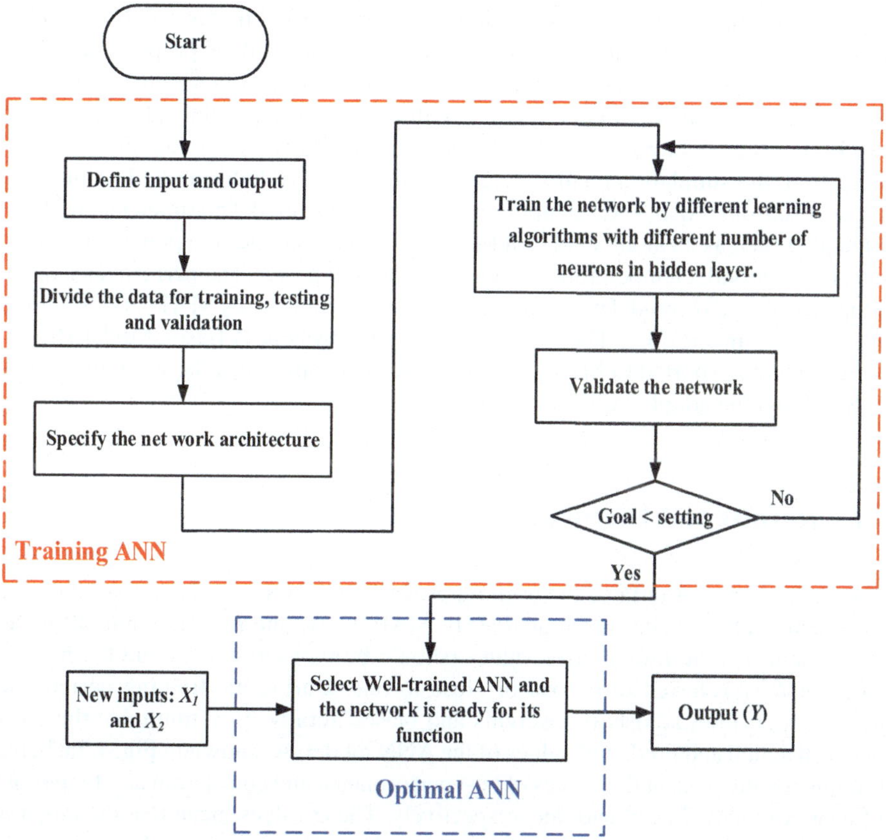


**Fig. 4.6** Flow chart of ANN training processes

- Develop an ANN model for analysis and define the input and output parameters.
- Normalization of input and outputs either in the range between 0 and 1 or between − 1 and 1 depending on the type of data and transfer function used.
- Train the model with normalized input and output data by different learning algorithms with different number of neurons in hidden layer.
- Calculate the performance of different models on the basis of statistical error analysis.
- Selection of the optimal ANN model on the basis of statistical performance of predicted data with higher values of coefficient of determination and lower values of different errors.
- Now the ANN model is ready for its function.

## 4.5 Validation and Analysis

This section shows and discusses the simulation results of CTR-ANN model as explained in chapters three and four. In this analysis, RBF, MLP, and GRNN models were applied for estimating HTF mass flow rate. For training purpose, the three models were examined with an increased number of spread factor and neurons to determine the output accurately. As well as, three learning algorithms; LM, BFG, and SCG, were selected for training MLP model. Additionally, for evaluating the model validation, the simulation results were performed over 70 h during both winter and summer seasons. Also, several inputs and outputs of the model parameters have been simulated such as the hourly solar radiation, receiver solar thermal power, mass flow rate, receiver outlet temperature, and electrical output power. The performance of the adopted CTR-ANN model was analysed and implemented using MATLAB/Simulink as described in Fig. 3.2. The obtained results by System Advisor Model (SAM) software were exported to MATLAB® to perform a comparison between the results of SAM and the adopted CTR-ANN model.

## 4.6 Results and Discussion

The performance criteria of the training process for ANN models are based on the statistical analysis values such as RMSE, MAE, COV, and $R^2$. The optimal model is that achieves the lowest error values (should be 0 or close to 0) and the highest value of $R^2$ (preferred to be one or close to one. During the training period, the training algorithm adjusts the weights and biases iteratively to minimize the error between actual and predicted values of the ANN model. As shown in Fig. 4.7a, b, the training results present the values of best performance and correlation coefficient ($R$) of approximately 7.2e$^{-5}$ and one, respectively. These values mean that the target is equal to the output of the training data, which clarify the strong correlation between

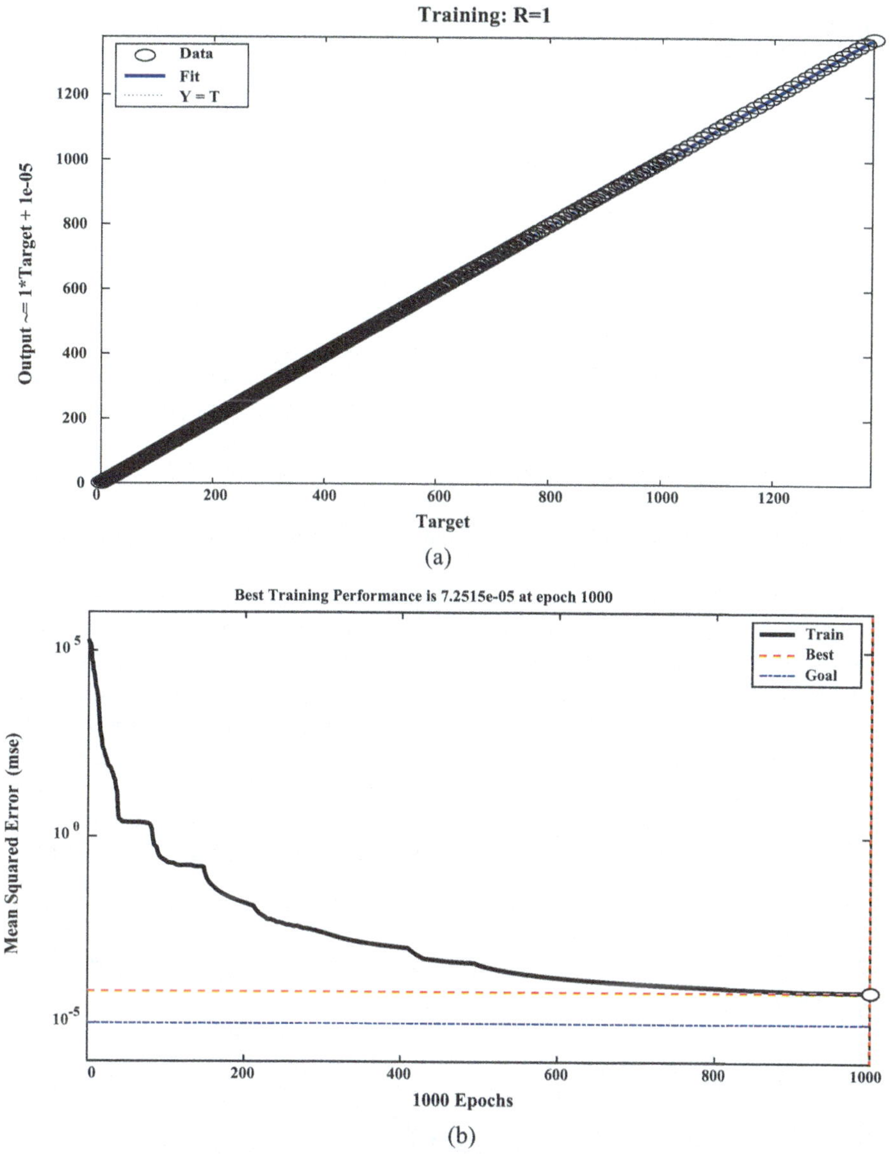


**Fig. 4.7** Training results of ANN model: **a** Regression plot and **b** Model performance

the target data and the ANN output. Hence, it can be confirmed that the proposed ANN model is of high accuracy and can successfully predict the desired target data.

In order to determine the optimal model of ANN technique, the three ANN models were trained at different spread factors (from 0.1 to 2 for GRNN and from 1.2 to 4.3 for RBF) and by varying the number of neurons in the hidden layer (from 5 to

**Table 4.2** Performance evaluation of GRNN model

| Spread factor | RMSE | MAE | $R^2$ |
|---|---|---|---|
| 0.1 | 1.8346 | 0.8874 | 0.9850 |
| 0.5 | 1.8332 | 0.8868 | 0.9850 |
| 0.7 | 1.7567 | 0.8502 | 0.9862 |
| 1 | 1.2242 | 0.6044 | 0.9931 |
| 1.5 | 0.4817 | 0.32 | 0.9990 |
| 1.8 | 0.6577 | 0.3521 | 0.9990 |
| 2 | 0.8127 | 0.402 | 0.9989 |

40 for MLP) as summarized in Tables 4.2 to 4.4. This procedure is called trial and error method that is most common method for determining the optimal number of neurons in the hidden layer or spread factor. The main constraint of trial and error method is that this method is extremely time-consuming. About 70% of the total dataset is used to form the weight factors and bias in the training process, while the remainder is used in the testing process to calculate the prediction error for estimating the accuracy of the models.

Tables 4.2 to 4.4 show the predicted results of the performance evaluation based on statistical errors analysis for the testing dataset for GRNN, RBF, and MLP models, respectively. Then RMSE, MAE and $R^2$ are computed and the performance index of ANN model are given in these tables. Also to assess the accuracy of ANN estimations, the regression analysis between the network response (predicted outputs) and the corresponding targets are shown in Figs. 4.8, 4.9, 4.10. Therefore, ANN's estimation values of discharge rate were assessed with regression analysis between estimated and calculated data. When a perfect fit had been found (predicted outputs were exactly the same as the calculated data), the slope of this straight $R^2$ line would be

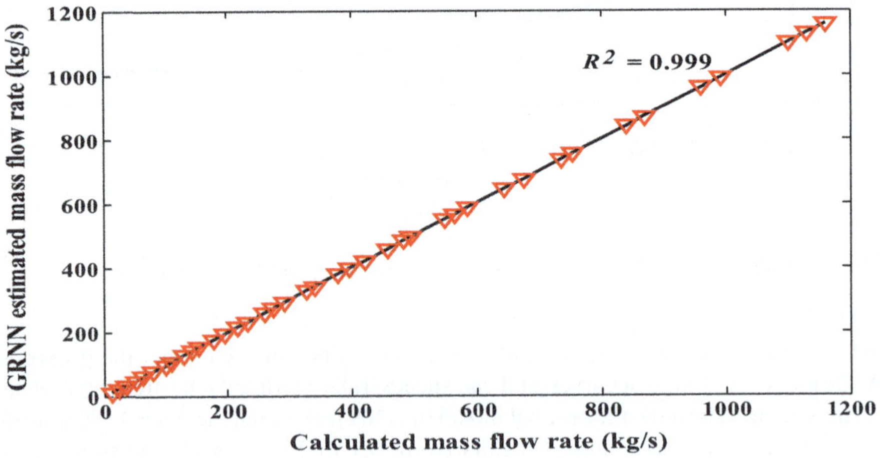


**Fig. 4.8** Calculated mass flow rate and the estimated by GRNN model

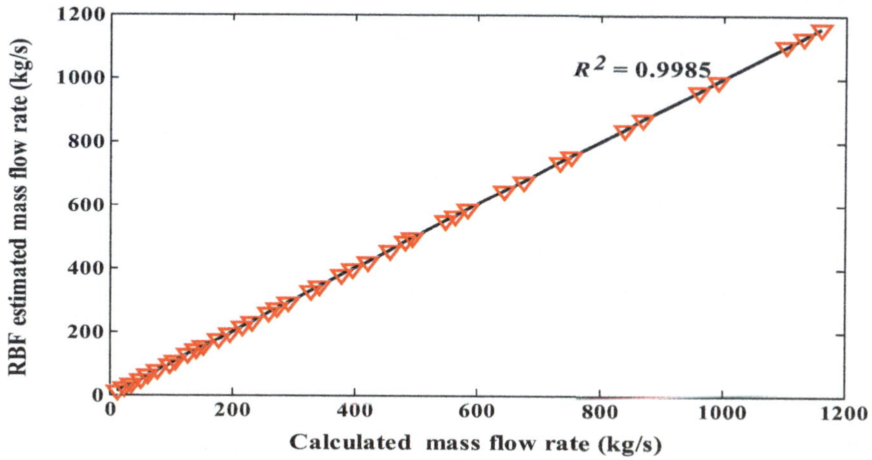


**Fig. 4.9** Calculated mass flow rate and the estimated by RBF model

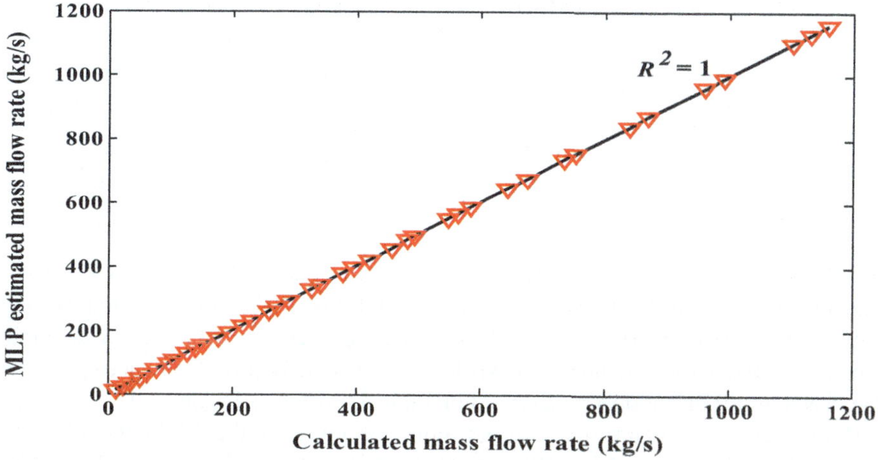


**Fig. 4.10** Calculated mass flow rate and the estimated by MLP model

one. Hence, Figs. 4.8 to 4.10 show that these values are very close, which indicates a good response to training and testing sets.

For GRNN model at a spread factor of 1.5 as shown in Table 4.2, the best value of $R^2$ was 0.999 and the lowest values of RMSE and MAE were 0.4817 and 0.32, respectively. Similarly, RBF model presents its best performance at a spread factor of 4 as demonstrated in Table 4.3; the best value of $R^2$ was 0.9985 and the lowest values of RMSE and MAE were 0.2846 and 0.0674, respectively. While in the case of using MLP model as exhibited in Table 4.4, it is found that the highest value of $R^2$ was one at neurons number of 40 and the lowest RMSE and MAE were 0.003 and

**Table 4.3** Performance evaluation of RBF model

| Spread factor | RMSE | MAE | $R^2$ |
|---|---|---|---|
| 1.2 | 62.4509 | 24.3655 | 0.1848 |
| 1.7 | 7.8996 | 3.0197 | 0.9894 |
| 2 | 2.6198 | 0.7555 | 0.9996 |
| 2.3 | 2.3529 | 0.4298 | 0.9999 |
| 2.5 | 2.2988 | 0.3994 | 0.9999 |
| 3 | 1.2406 | 0.23184 | 1 |
| 3.5 | 0.4539 | 0.0944 | 0.9829 |
| 4 | 0.2846 | 0.0674 | 0.9985 |
| 4.3 | 0.353 | 0.0616 | 1 |

**Table 4.4** Performance evaluation of MLP model

| Neurons number | RMSE | MAE | $R^2$ |
|---|---|---|---|
| LM-40 | 0.003 | 0.0023 | 1 |
| LM-35 | 0.0062 | 0.0056 | 1 |
| LM-30 | 0.0059 | 0.0049 | 1 |
| LM-25 | 0.0162 | 0.0125 | 1 |
| LM-20 | 0.0105 | 0.0082 | 1 |
| LM-15 | 0.0221 | 0.0171 | 0.9999 |
| LM-10 | 0.0174 | 0.0145 | 1 |
| LM-5 | 0.0313 | 0.0231 | 0.9998 |

0.0023, respectively. Furthermore, as presented in Figs. 4.11, 4.12, to 4.13, it was seen that most of the absolute relative errors for GRNN, RBF, and MLP models are in between 0 – 0.018, 0 – 0.0325, and 0 – 0.00015, respectively, which are acceptable. Among all, it can be concluded that MLP model has the best performance compared to GRNN and RBF models as clarified in Table 4.4, Figs. 4.10, and 4.13.

Furthermore, to determine the optimum topology of the MLP model, three learning algorithms (i.e., LM, BFG, and SCG) were selected for training this model. The results of the model performance and regression plot for training, validation, testing and all process for the three training algorithms at neurons number of 40 in the hidden layer (i.e., LM-40, SCG-40, BFG-40) are described in Figs. 4.14a, b to 4.16a, b, respectively. The best MSE value for LM-40 was found 1.43E-5 at epoch 28 and the obtained values for $R$ were 0.9999, 0.99999, 0.99999, and 0.99999 for training, validation, testing, and all, respectively. Similarly, in the case of using SCG-40, the best MSE value was found 0.0049 at epoch 106, while $R$ values for training, validation, testing, and all were 0.99708, 0.9968, 0.9979, and 0.99717, respectively. Also, when BFG-40 was applied, the observed value for MSE was found 0.00119 at epoch 79 and $R$ values were 0.9993, 0.9991, 0.9987, and 0.9992 for training, validation, testing, and all, respectively (Fig. 4.15).

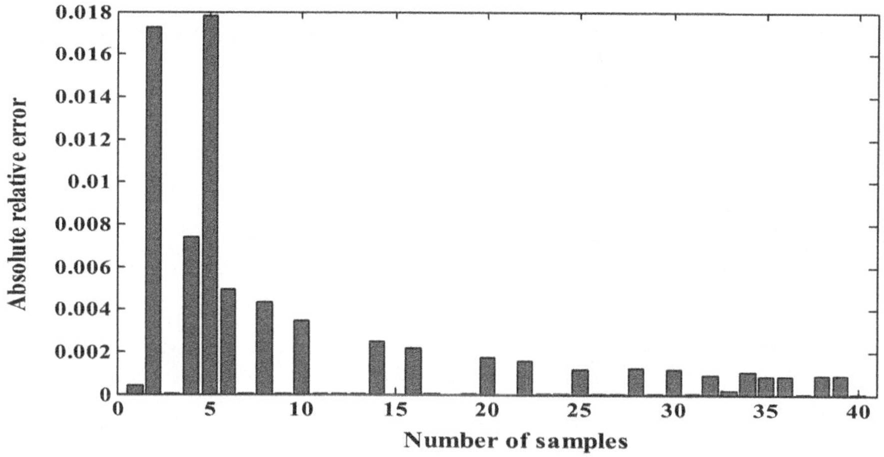


**Fig. 4.11** Relative error of GRNN model at data samples

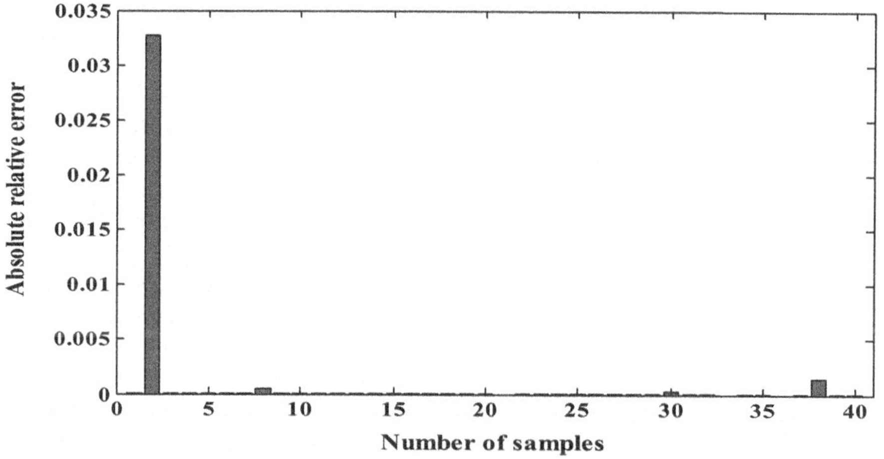


**Fig. 4.12** Relative error of RBF model at data samples

Table 4.5 illustrates the statistical results for training and testing data for LM, SCG, and BFG algorithms with several numbers of neurons from 5–40. As shown in Table 4.5, the training algorithm LM-40 displays the lowest values of RMSE (0.0032) and COV (0.0011) at the training process and RMSE (0.0041) and COV (0.0015) at the testing process. While, the values of $R^2$ were 0.99999 and 0.99998 at the training and testing process, respectively. It can be concluded from the above analysis that the LM algorithm with 40 neurons is very satisfactory and the optimum topology compared with CSG and BFG algorithms, which is ascribed to the higher values of $R^2$ and the lowest values of both RMSE and COV. This implies that the LM algorithm

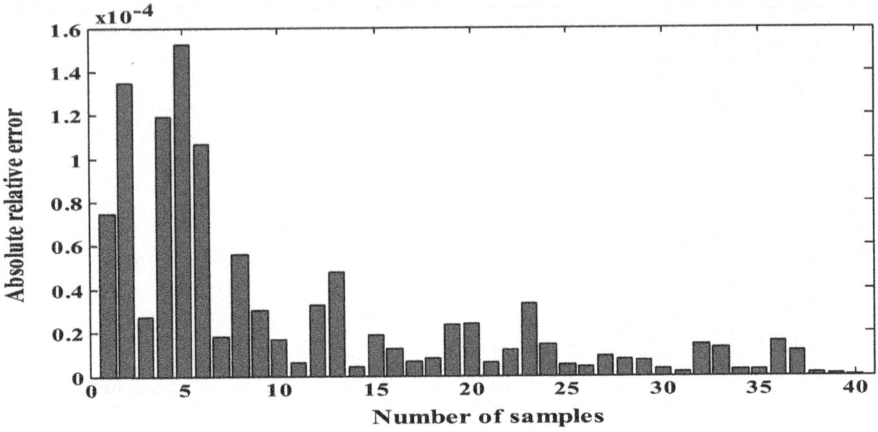


**Fig. 4.13** Relative error of MLP model at data samples

performs well and can be accurately follow the data patterns. Therefore, the MLP with LM-40 represents the best model that can be used with a suitable accuracy to estimate the discharge rate of the adopted CTR plant model.

Figure 4.17 shows a comparison between the calculated HTF mass flow rate from the CST to the tower receiver and the estimated discharge rate by the best MLP-LM model, considering the changes in the input parameters, i.e. receiver inlet temperature and thermal power. It is obvious that the predictive outputs of the MLP-LM model are identical with the target outputs, hence, it can accurately determine the optimal HTF mass flow rate to the tower receiver.

The performance of the adopted CTR-ANN model was compared with reference results by using SAM software in order to evaluate the model validation. The simulation results were performed for about three days during different year's seasons. Figure 4.18a, b shows the comparison between the calculated and simulated hourly solar radiation by using DI method and SAM software during winter and summer days, respectively. It is found that a slight difference between the two methods particularly during summer season. DI method result is slightly larger than SAM result. This inconsiderable difference between the two methods is a result of the meteorological data and solar resource in SAM weather file. Indeed, this file contains typical data for one year that may be obtained from satellite, ground measurements, or a combination of the two. Similarly, thermal powers of the adopted model and SAM are compared in Fig. 4.19a, b. In this simulation, the receiver efficiency of the adopted model was set at 90%. As well, thermal power of the adopted model is slightly larger than SAM result during summer season due to the calculated higher solar radiation by DI method as explained in Fig. 4.18b.

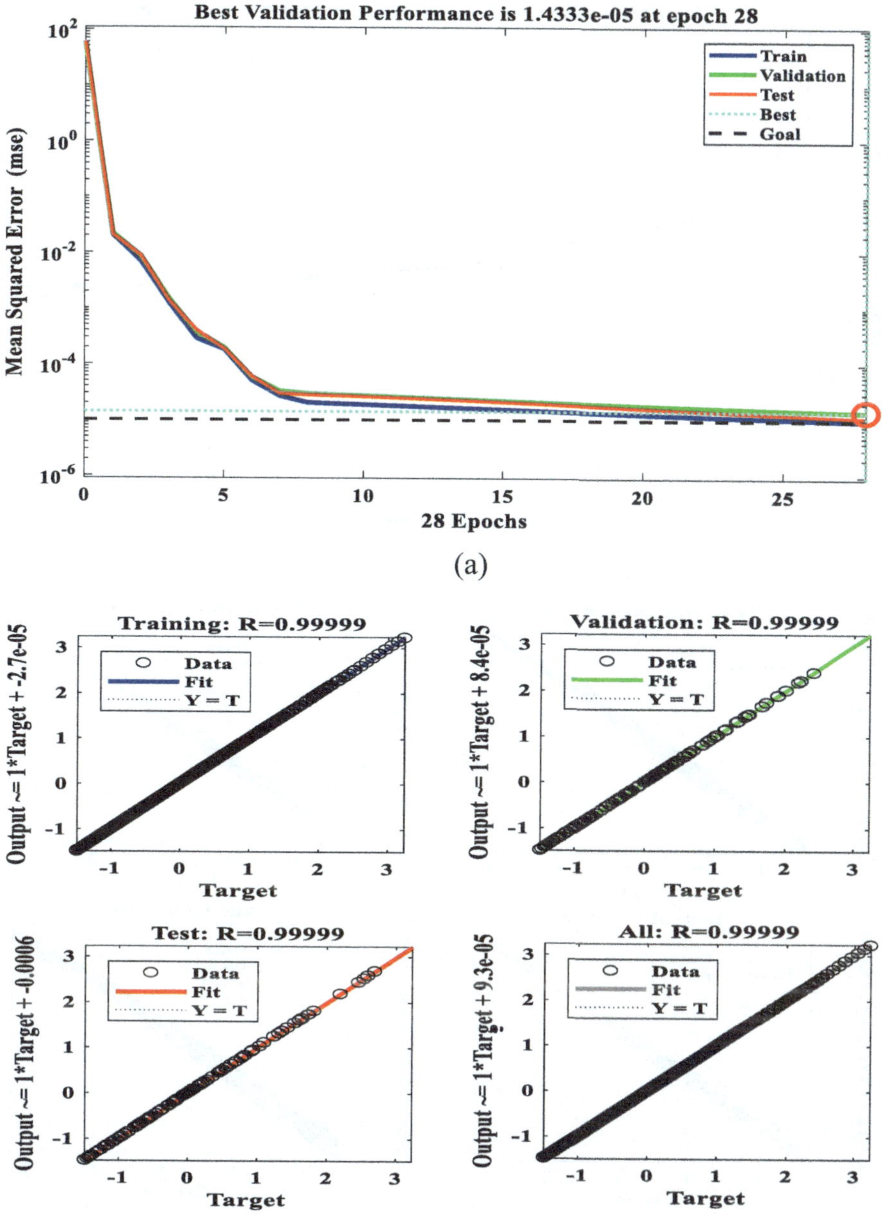


(a)

(b)

**Fig. 4.14** LM-40 algorithm: **a** Best performance and **b** Regression plots

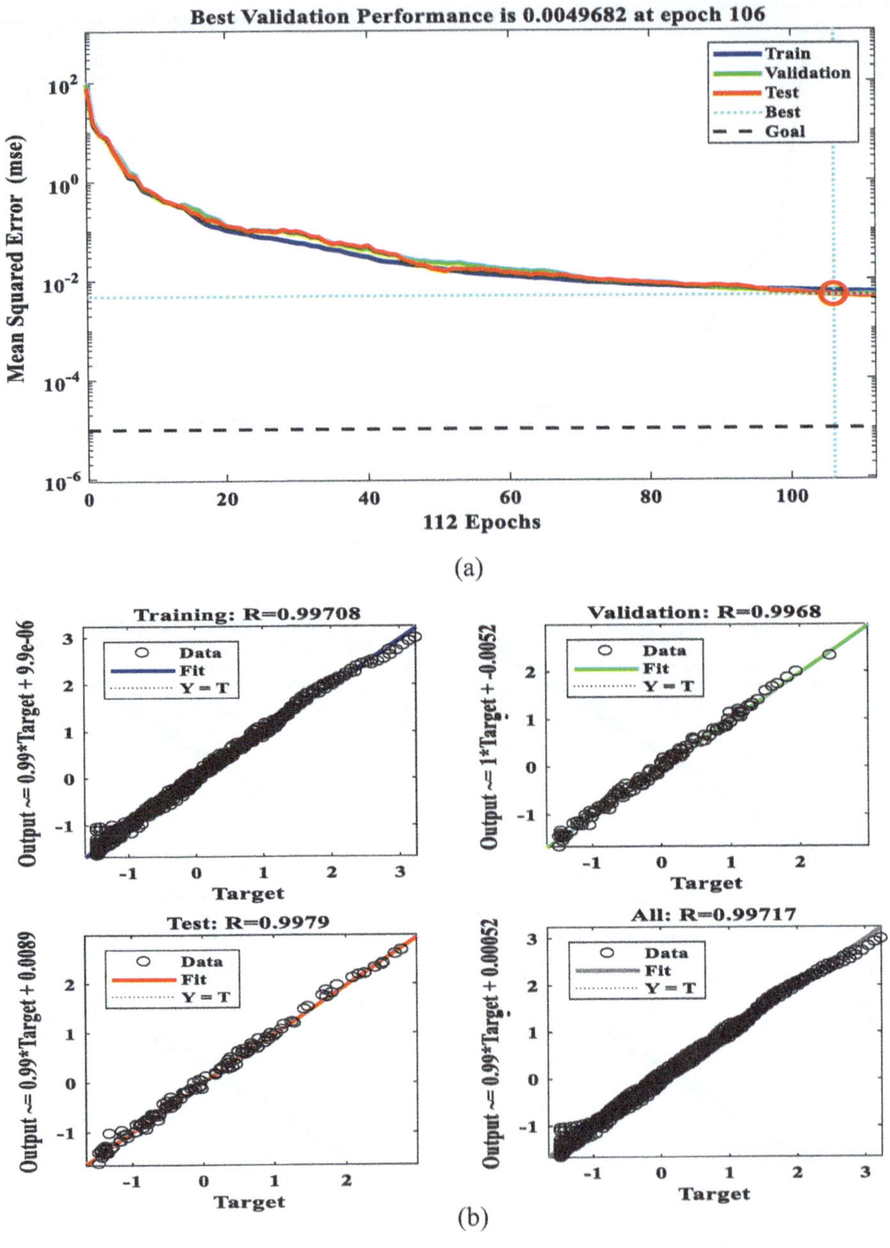


**Fig. 4.15** SCG-40 algorithm: **a** Best performance and **b** Regression plots

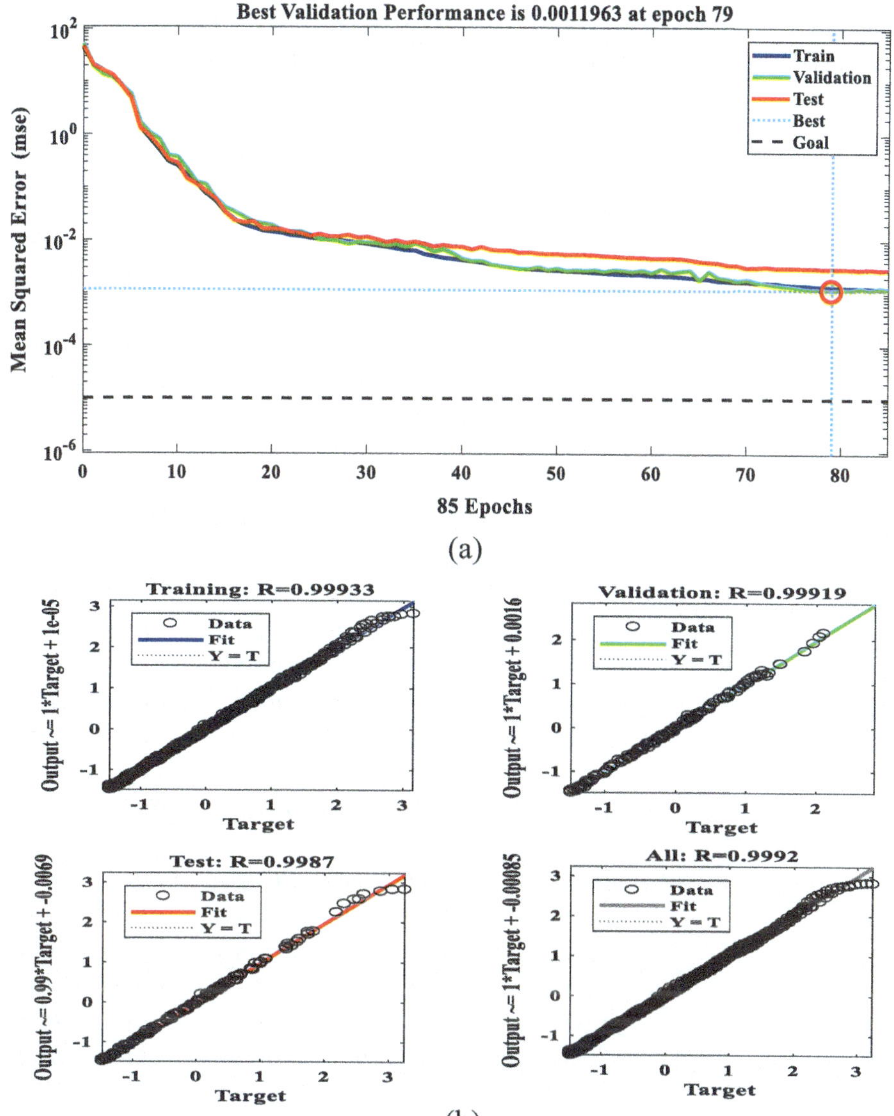


**Fig. 4.16** BFG-40 algorithm: **a** Best performance and **b** Regression plots

Furthermore, the mass flow rate, receiver outlet temperature, and electrical output power for the adopted model and SAM during two seasons of the year are presented in Figs. 4.20a–c and 4.21a–c, respectively. An inconsiderable deviation between the results of adopted model and SAM, which depends on an existence of a slight difference between the adopted and simulated thermal power, different control strategy, and

**Table 4.5** Comparison of error analysis of training and testing data for MLPNN topologies

| Algorithm and neurons No | Training data | | | Testing data | | |
|---|---|---|---|---|---|---|
| | RMSE | COV | $R^2$ | RMSE | COV | $R^2$ |
| LM | | | | | | |
| 5 | 0.0315 | 0.0113 | 0.9999 | 0.0323 | 0.0115 | 0.9999 |
| 10 | 0.0137 | 0.0049 | 0.9999 | 0.0130 | 0.0046 | 0.9999 |
| 15 | 0.0268 | 0.0096 | 0.9999 | 0.0283 | 0.0101 | 0.9999 |
| 20 | 0.0124 | 0.0044 | 0.9999 | 0.0137 | 0.0049 | 0.9999 |
| 25 | 0.0131 | 0.0047 | 0.9999 | 0.0164 | 0.0059 | 0.9999 |
| 30 | 0.0057 | 0.0020 | 0.9999 | 0.0056 | 0.0020 | 0.9999 |
| 35 | 0.0055 | 0.0013 | 0.9999 | 0.0059 | 0.0016 | 0.9999 |
| 40 | 0.0037 | 0.0011 | 0.9999 | 0.0041 | 0.0015 | 0.9999 |
| SCG | | | | | | |
| 5 | 0.0532 | 0.0190 | 0.9976 | 0.0542 | 0.0193 | 0.9967 |
| 10 | 0.0433 | 0.0155 | 0.9977 | 0.0425 | 0.0151 | 0.9980 |
| 15 | 0.0565 | 0.0202 | 0.9946 | 0.0630 | 0.0225 | 0.9948 |
| 20 | 0.0698 | 0.0249 | 0.9952 | 0.0770 | 0.0275 | 0.9960 |
| 25 | 0.0953 | 0.0340 | 0.9940 | 0.0883 | 0.0315 | 0.9928 |
| 30 | 0.0749 | 0.0267 | 0.9936 | 0.0751 | 0.0268 | 0.9943 |
| 35 | 0.0809 | 0.0289 | 0.9930 | 0.0780 | 0.0278 | 0.9931 |
| 40 | 0.0705 | 0.0285 | 0.9942 | 0.0703 | 0.0280 | 0.9951 |
| BFG | | | | | | |
| 5 | 0.0285 | 0.0102 | 0.9990 | 0.0314 | 0.0112 | 0.9989 |
| 10 | 0.0211 | 0.0075 | 0.9995 | 0.0234 | 0.0083 | 0.9994 |
| 15 | 0.0329 | 0.0117 | 0.9991 | 0.0347 | 0.0124 | 0.9990 |
| 20 | 0.0136 | 0.0049 | 0.9992 | 0.0141 | 0.0050 | 0.9987 |
| 25 | 0.0302 | 0.0107 | 0.9991 | 0.0278 | 0.0099 | 0.9984 |
| 30 | 0.0346 | 0.0124 | 0.9973 | 0.0364 | 0.0130 | 0.9986 |
| 35 | 0.0311 | 0.0111 | 0.9991 | 0.0305 | 0.0109 | 0.9986 |
| 40 | 0.0555 | 0.0199 | 0.9983 | 0.0345 | 0.0123 | 0.9977 |

different operating conditions. Discharge rate from CST to tower receiver proportional to the available radiation during different seasons of the year as shown in Figs. 4.20a and 4.21a. It is clear that the simulated mass flow rate by MLP neural network and SAM are identical. However, there is a small difference at the beginning of discharge owing to the difference in the minimum value of thermal power of the two methods as explained in chapter three. It is also observed in Figs. 4.20b and 4.21b that the receiver outlet temperature is constant during the operating time of solar power plant because the discharge rate from CST varies with thermal power at the receiver.

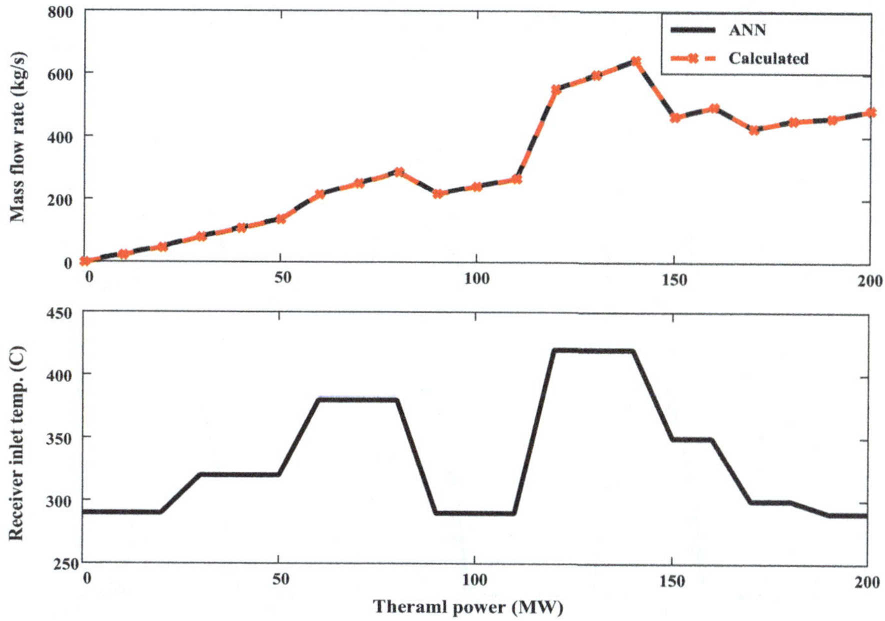


**Fig. 4.17** Comparison between calculated mass flow rate and estimated by ANN

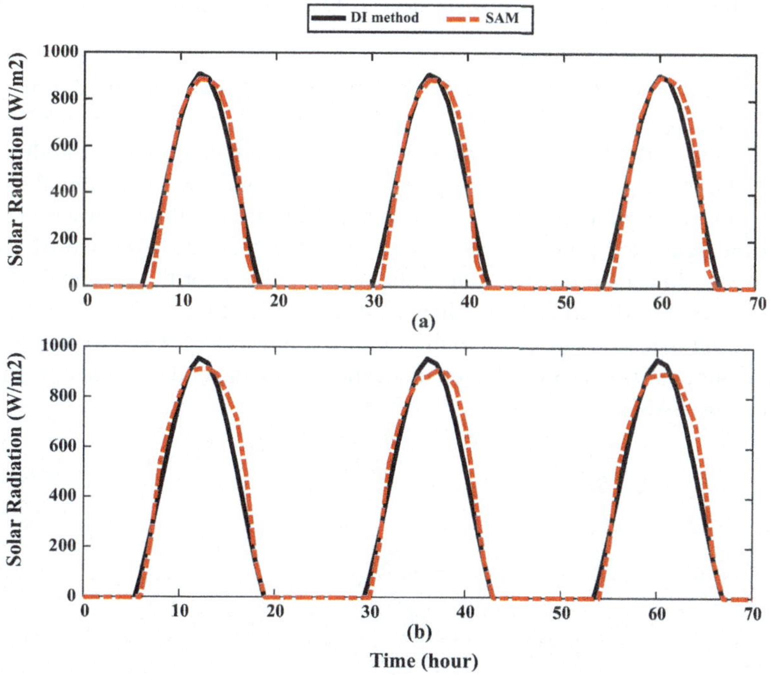


**Fig. 4.18** Hourly solar radiation: **a** Winter days (1–3 January) and **b** Summer days (21–23 June)

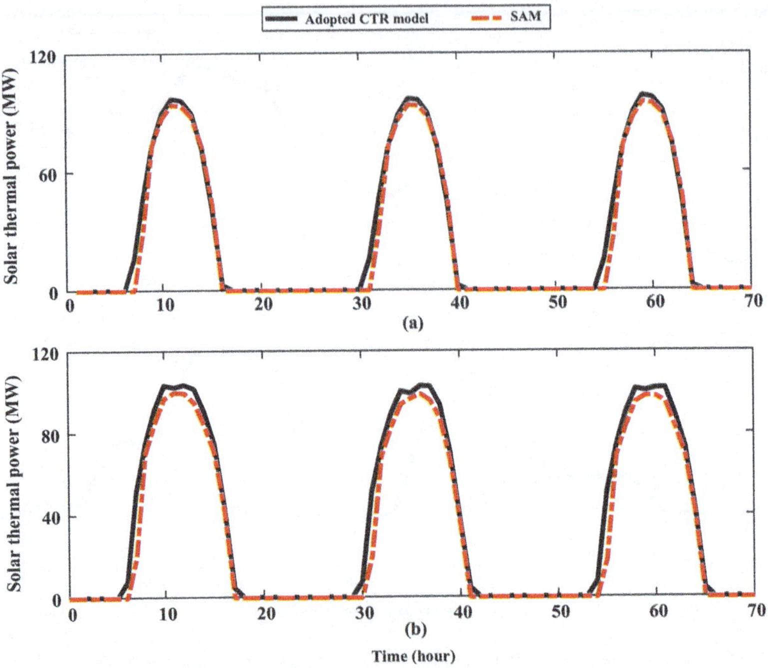


**Fig. 4.19** Receiver thermal power: **a** Winter days (1–3 January) and **b** Summer days (21–23 June)

Generator output of adopted CTR-ANN model and SAM software during winter and summer days are described in Figs. 4.20c and 4.21c, respectively. SAM output power is not regular because its strategy depends on the conditions of both SF and TES. While in the present work, the power has been regulated by making the output directly relies on HST conditions as explained in chapter three. However, if the area of the generated power curve is calculated, the energies harvested by the two methods are almost similar. Moreover, the output powers for the two methods have been compared at different SM values as described in Fig. 4.22a–c. The adopted model output is reasonably close to the simulated power by SAM. Obviously, these comparisons confirm that the adopted model results are in good agreement with those simulated by SAM.

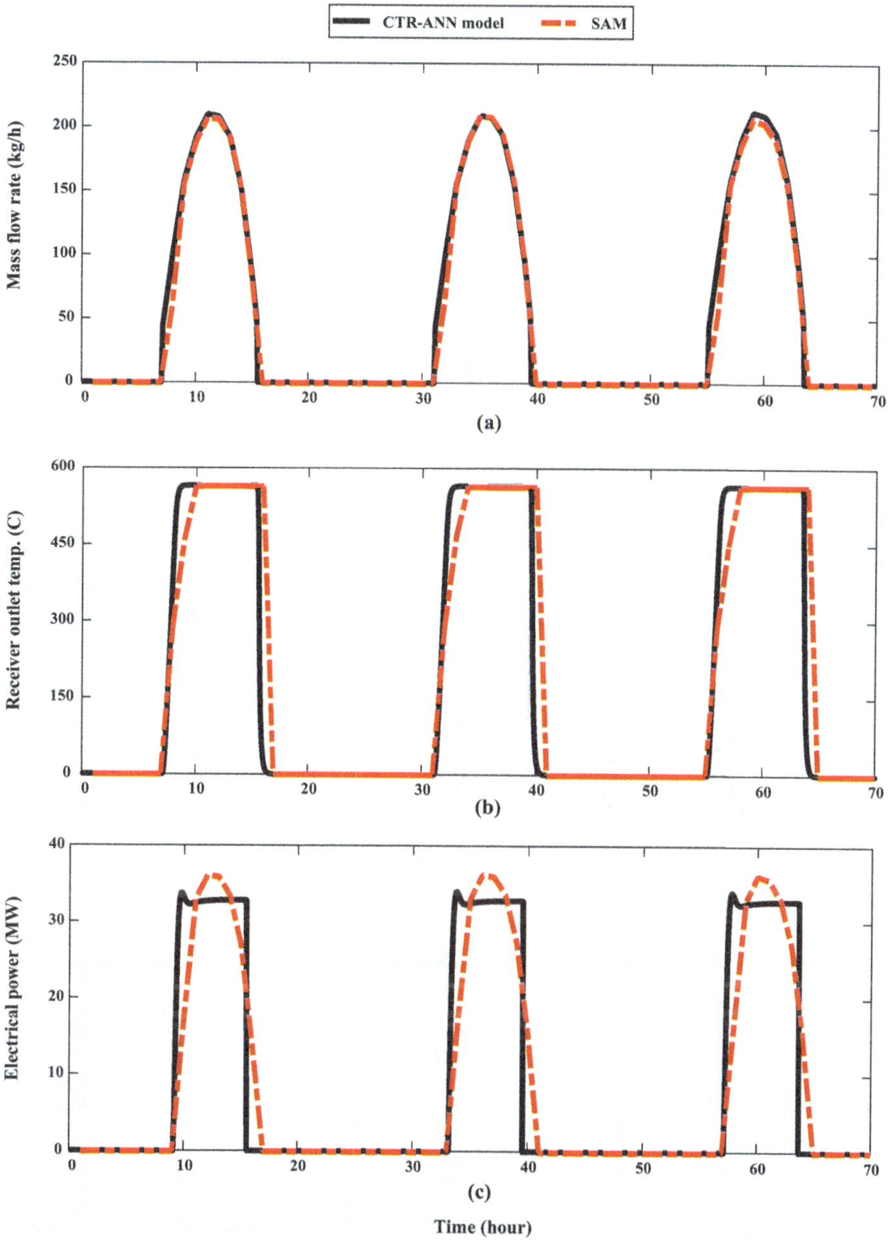


**Fig. 4.20** Comparison between adopted model and SAM during winter days: **a** Discharge rate, **b** Receiver outlet temperature, and **c** CTR output power

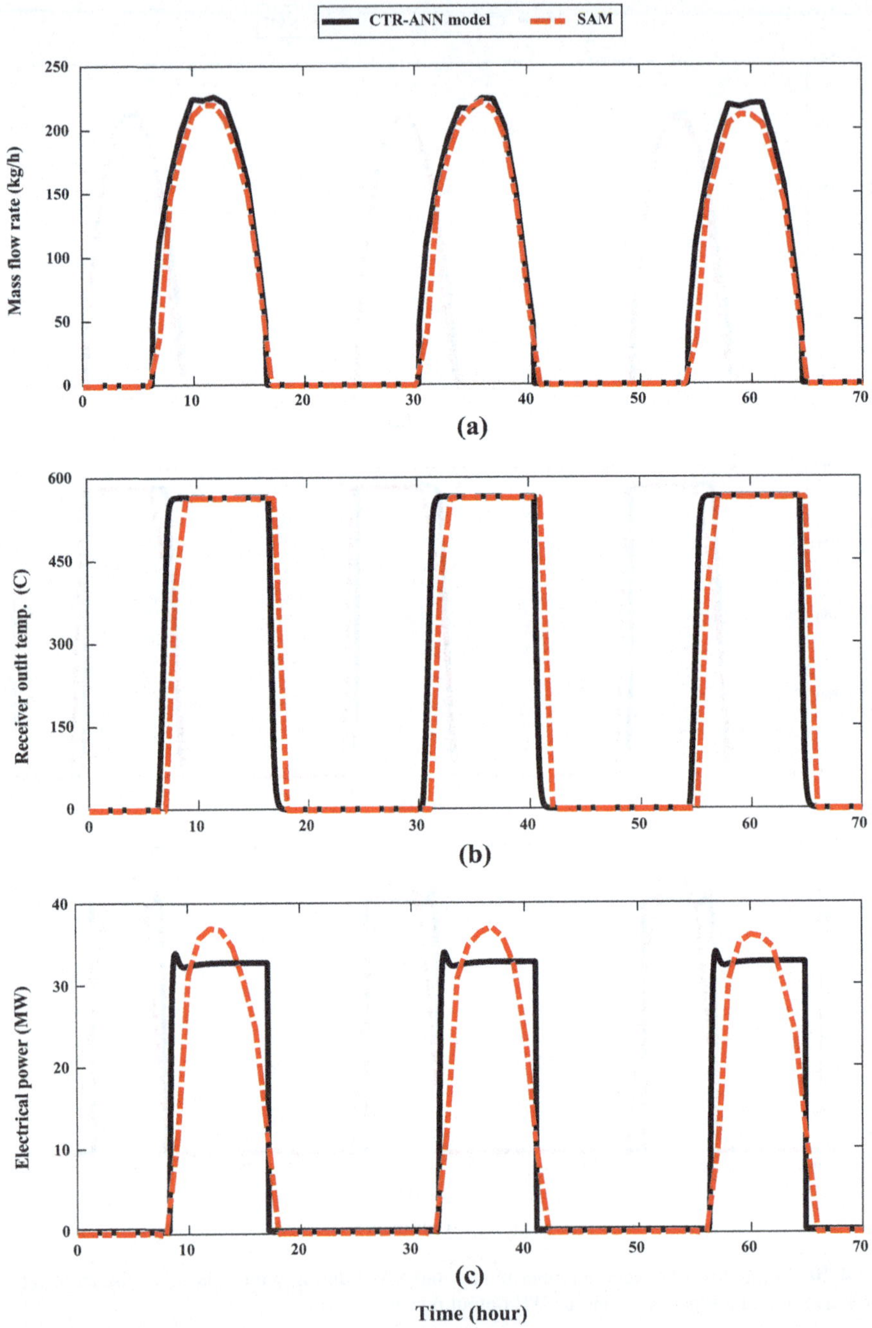


**Fig. 4.21** Comparison between adopted model and SAM during summer days: **a** Discharge rate, **b** Receiver outlet temperature, and **c** CTR output power

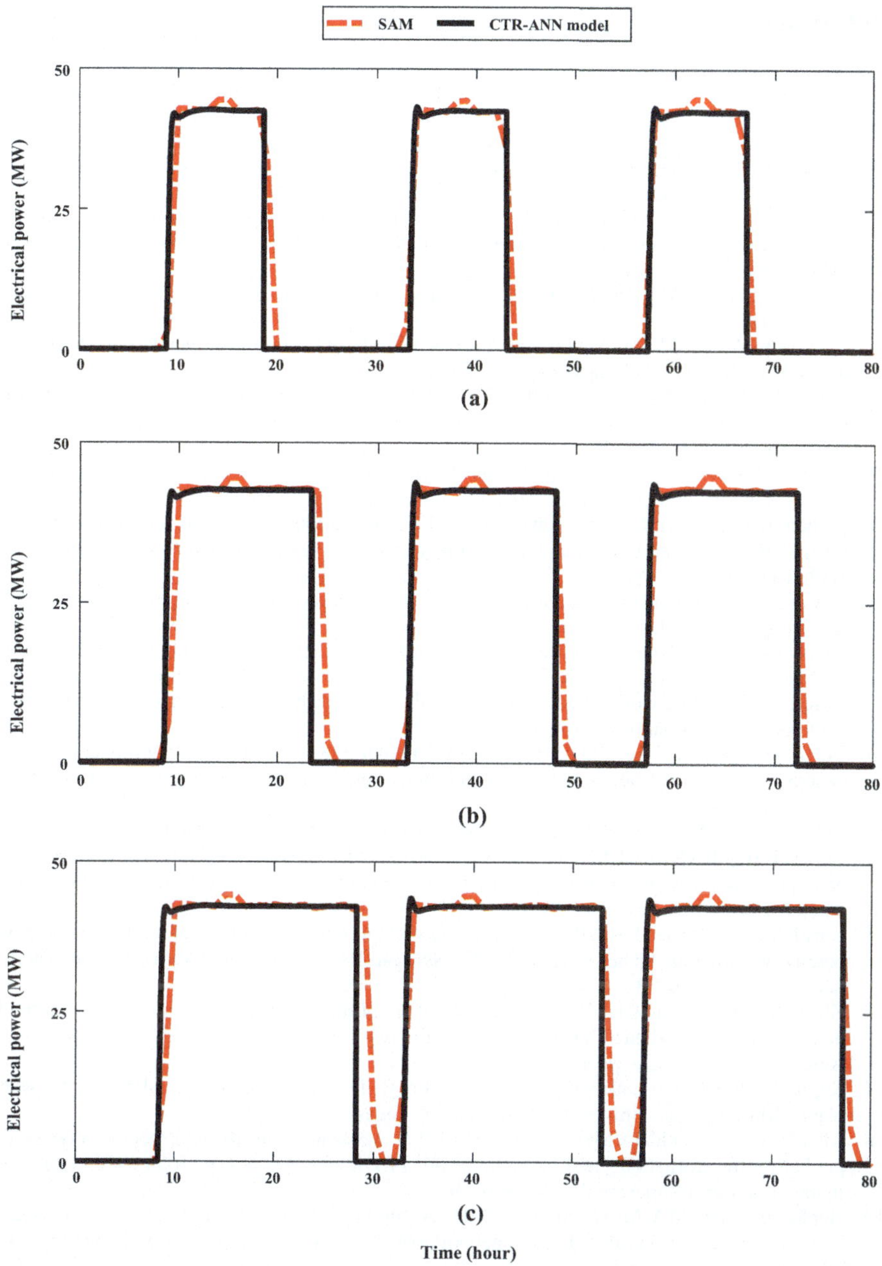


**Fig. 4.22** Comparison of adopted model and SAM power: **a** SM = 2, **b** SM = 3, and **c** SM = 4

## References


1. Haykin S (1994) Neural networks: a comprehensive foundation, 2nd edn. Prentice Hall PTR
2. Ghritlahre HK, Prasad RK (2018a) Application of ANN technique to predict the performance of solar collector systems-A review. Renew Sustain Energy Rev 84:75–88
3. Kerdphol T (2016) Optimization of battery energy storage systems for microgrids, PhD thesis. Graduate School of Engineering, Kyushu Institute of Technology, Japan
4. Moukhtar I et al (2017) A developed concentrated solar power model using artificial neural network technique. In: Nineteenth international middle east power systems conference (MEPCON'2017), pp 1346–1351
5. Beale MH, Hagan MT, Demuth H (1992) Neural network toolbox user's guide, 1st edn. Math Works, Inc
6. Chaouachi A et al (2009) Neural network ensemble-based solar power generation short-term forecasting. Int J Elect Comput Eng 3(6):54–59
7. Maier HR, Jain A, Dandy GC, Sudheer KP (2010) Methods used for the development of neural networks for the prediction of water resource variables in river systems: Current status and future directions. Environ Model Softw 25(8):891–909
8. Azizi S, Ahmadloo E (2016) Prediction of heat transfer coefficient during condensation of R134a in inclined tubes using artificial neural network. Appl Therm Eng 106:203–210
9. Mohanraj M, Jayaraj S, Muraleedharan C (2012) Applications of artificial neural networks for refrigeration, air-conditioning and heat pump systems—a review. Renew Sustain Energy Rev 16(2):1340–1358
10. Faucett L (1994) Fundamentals of neural networks: Arcihitectures, Algorithms, and Applications. Prentice Hall-Inc., Canada
11. Specht DF (1991) A general regression neural network. IEEE Trans Neural Net 2(6):568–576
12. Al-Mahasneh AJ, Anavatti SG, Garratt MAJ (2018) Review of applications of generalized regression neural networks in identification and control of dynamic systems. In: International conference on aeronautics, astronautics and aviation
13. Li HZ, Guo S, Li CJ, Sun JQ (2013) A hybrid annual power load forecasting model based on generalized regression neural network with fruit fly optimization algorithm. Knowl-Based Syst 37:378–387
14. Irfan M, Koj A, Thomas H, Sedighi M (2016)Geographical General Regression Neural Network (GGRNN) tool for geographically weighted regression analysis. In: Eighth international conference on advanced geographic information systems, applications, and services, pp 154–159
15. Ghritlahre HK, Prasad RK (2018b) Investigation of thermal performance of unidirectional flow porous bed solar air heater using MLP, GRNN, and RBF models of ANN technique. Therm Sci Eng Prog 6:226–235
16. Ghritlahre HK, Prasad RK (2018) Development of optimal ANN model to estimate the thermal performance of roughened solar air heater using two different learning algorithms. Springer, Annals of Data Science, pp 1–15
17. Kopp JE (2009) Two-tank indirect thermal storage designs for solar parabolic trough power plants, Master thesis, University of Nevada Las Vegas
18. Khajeh M, Moghaddam MG, Shakeri M (2012) Application of artificial neural network in predicting the extraction yield of essential oils of Diplotaenia cachrydifolia by supercritical fluid extraction. J Supercrit Fluids 69:91–96
19. Heidari E, Sobati MA, Movahedirad S (2016) Accurate prediction of nanofluid viscosity using a multilayer perceptron artificial neural network (MLP-ANN). Chemom Intell Lab Syst 155:73–85
20. Aghbashlo M, Mobli H, Rafiee S, Madadlou A (2012) The use of artificial neural network to predict exergetic performance of spray drying process: A preliminary study. Comput Electron Agric 88:32–43

# Chapter 5
# Penetration Characteristics of Hybrid CSP and PV Solar Plants Economic

**Abstract** The rapid increase in the integration of PV system over the past decades is due to its falling cost and advantages. High penetration of solar PV system causes a significant effect on the power quality of the system due to the mismatch between demand patterns and the solar resource. The negative impact of PV penetration on the system limits its high penetration. Therefore, PV itself may represent a new challenge to the electrical system rather than being a part of the solution. Hence, the CSP technology including TES has been aggregated in the system in order to accommodate the required electrical power during the higher and lower solar energy at all timescales. Therefore, this chapter studies the CSP impacts on the penetration level of PV system as well as on the reliability of the system in two cases: the first case is only integration of PV system. The second case by using hybrid PV and CSP systems. The results showed that the performance of CSP technologies has a significant positive impact on the system, which supports the overall flexibility of the system because of its ability to store and send generated energy. In addition, the use of CSP reduces the minimum generation constraint of the conventional generators that allows more penetration of the PV system.

**Keywords** Photovoltaic (PV) · Concentrated solar power (CSP) · Thermal energy storage (TES) · Penetration level · Conventional generators · Overall flexibility · Generation constraint

## 5.1 Introduction

The falling cost and advantages of RES, especially solar PV system, led to a rapid increase in the integration of PV system over the past decades. Although, the high penetration of solar PV system has a significant effect on the power quality of the system due to the mismatch between demand patterns and the solar resource [1, 2]. The high penetration problems of PV system increase when the sun radiation is low during the periods of cloudy weather and after sunset. The negative impact of PV penetration on the system limits its high penetration. The severity of these effects depends on the degree of penetration of solar PV system [3]. Therefore, PV itself may represent a new challenge to the electrical system rather than being a part of the



I. Moukhtar et al., *Solar Energy*, Power Systems,
https://doi.org/10.1007/978-3-030-61307-5_5

solution. Although, batteries utilization is a solution for the high penetration problems of solar PV system, but their cost and lifetime are the great financial challenges facing their wide application.

Hence, the CSP technology including TES has been aggregated in the system in order to accommodate the required electrical power during the higher and lower solar energy at all timescales. TES is a storage media that stores the excess power and releases it during overcast or nighttime periods. Incorporating TES into CSP plants increases the number of operation hours, system energy efficiency, and PB utilization and reduces the times of mismatch between energy demand and energy supply by the sun; this leads to make the integration of solar systems easier [4]. This work analyzes the impacts of CSP/TES technology integration on penetration level of PV system, particularly when no energy storages in the form of batteries are used. Where CSP/TES is considered as a better solution for the high penetration problem of solar energy, due to its ability to store and dispatch the power depending on the load demand [5]. This study analyzes the interconnection of solar energy with conventional generators for several configurations in order to be able to select the appropriate unit's size and maintain the balance between generation and demand at all timescales. This helps to utilize all the available solar energy and consequently reduces the required output from conventional generators. Generation sources have been scheduled to meet the load demand based on the data observation and forecasting of solar radiation for the location of Aswan city. The results show that the performance of CSP has a great impact on the penetration level of PV system and on the system flexibility. The overall flexibility increases due to the ability of CSP to store and dispatch the generated power. In addition, CSP/TES itself has inherent flexibility. Therefore, CSP reduces the minimum generation constraint of the conventional generators that allows more penetration of the PV system.

## 5.2 Methodology

### *5.2.1 System Description*

The major concern in the operation of an electrical system that utilizes RES is the accurate selection of components that can satisfy the load demand [6]. Figure 5.1 describes an example for hybrid solar energy systems. The model consists of local load and generation sources such as conventional generators and two types of solar energy sources. In this model, the conventional generators operate at a certain level of its capacity due to its ramping capability; 'Ramping capability means the generator's ability to decrease or increase the output power in order to follow and meet the demand behavior. When the power from RES cannot meet the load demand, the strategy will be used to find the economic solution, starting the conventional generator. Figure 5.2 shows the daily load curve that used in this study [7]. It is clear that the load demand increases continuously from the middle of the day until the end of the night and

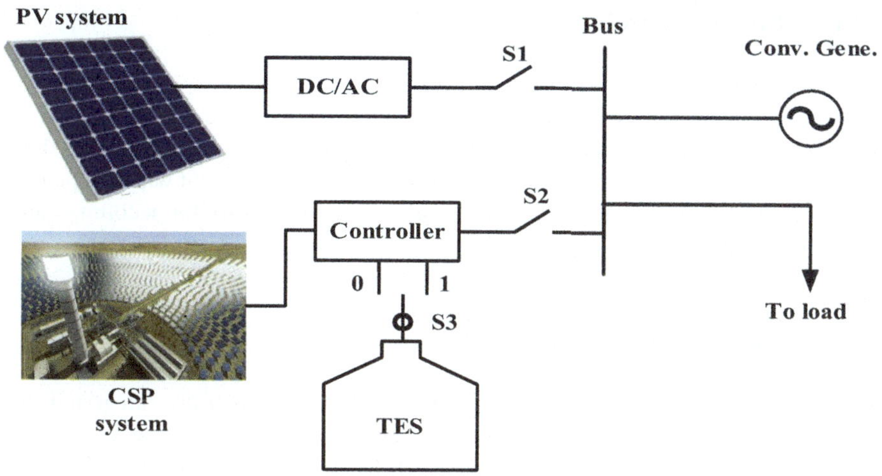


**Fig. 5.1** General configuration of the hybrid PV and CSP systems

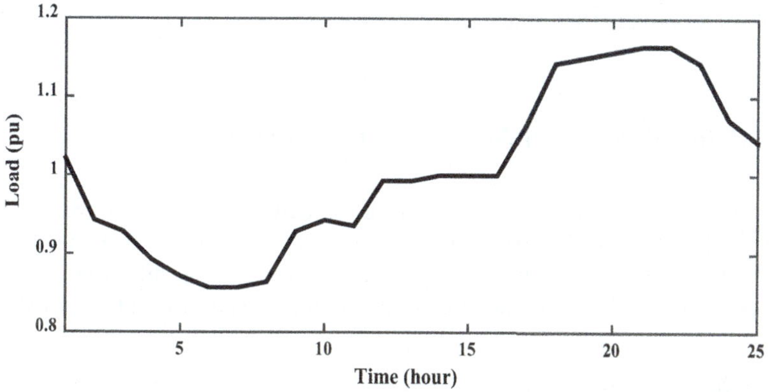


**Fig. 5.2** Daily load curve [7]

after that decreases again at the morning. Daily scale peak load is taken as 1.15 pu. It should be noted that the 1.15 pu peak load occurs at hour of 22 am. The largest demand occurs between hours of 16 pm and 24 pm and the lowest demand happens between hours of 1 am and 15 pm. The balance between generations and demand is necessary to maintain the power system reliability. Therefore, the electrical power from the conventional generators and RES must be adapted to the demand at all times periods along the day.

For analyzing CSP performance on the system for the location of Aswan city in Egypt, this site has a significant solar energy due to its clear and hot weather along the year especially during summer season, two cases of solar energy integration are discussed as below.

#### 5.2.1.1 Case 1: PV Only

This case study only one type of solar energy (solar PV system) is integrated with conventional generators to supply the load demand. To meet this case, switches S1 and S2 are in closed and open state, respectively, as shown in Fig. 5.1. This study shows the limits of PV penetration and displays the amount of curtailment which depending on the penetration level and the conventional generator's ability for accommodating the intermittent of solar energy.

#### 5.2.1.2 Case 2: Hybrid PV and CSP

In this case, two types of solar energy (hybrid PV and CSP) are integrated with conventional generators to feed the load demand. To achieve this case, switches S1 and S2 are in closed state as shown in Fig. 5.1. Switch S3 is used to store the excess power; S3 will be on state of charge in the position '0', while S3 will be on state of discharge in the position '1'. This case analyzes the advantage and performance of CSP on the system flexibility, how reduces the amount of PV dissipated (unused) and allows more penetration level.

### *5.2.2 Output Power of Solar Energy Sources*

System Advisor Model (SAM) software is developed by NREL to simulate the output power of RES such as PV, CSP, and wind energy [8, 9]. Figure 5.3 shows the output power of PV system that simulated by SAM software during different seasons. PV output power depends on weather conditions. At the first day, the output of PV system

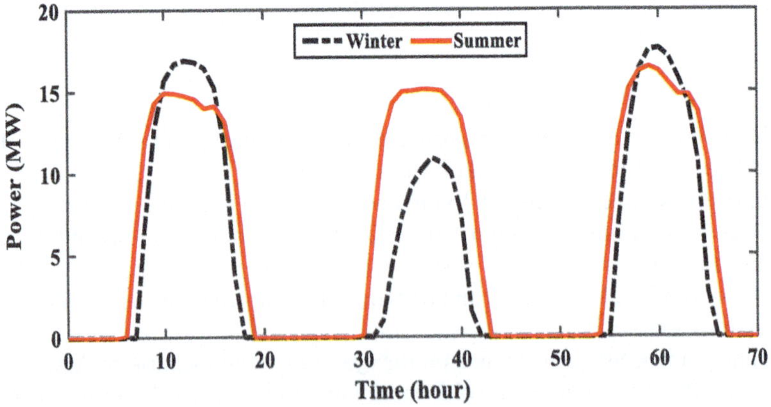


**Fig. 5.3** Solar PV output power during winter and summer seasons

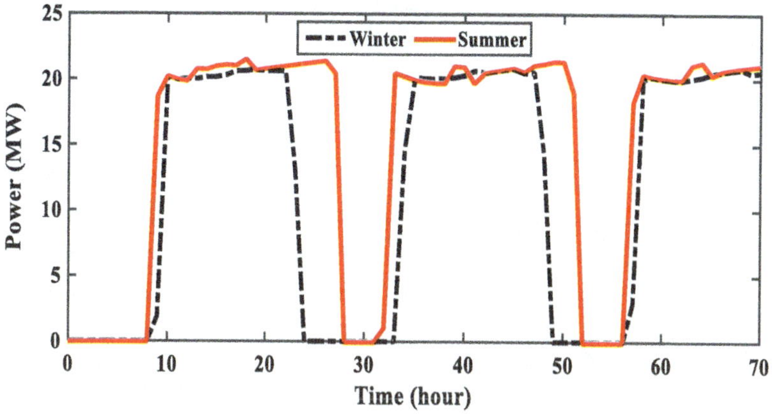


**Fig. 5.4** Solar CSP output power during winter and summer seasons

during the summer season is less than output power during the winter season due to the temperature effect as shown in Fig. 5.3. While PV output power in the second day is shrunk during the winter season due to the cloudy weather. During the winter season as shown in Fig. 5.4, it is observed that CSP plant has the ability to store and dispatch the electrical power to supply the demand along the entire day (from 7.5 to 24.5 h). While during the summer season, more thermal energy is stored in the HST. Consequently, CSP plant is more efficient and it can dispatch the electrical power for a long time covering the night period of the first day and reaches the next day (from 7.5 to 28.5 h).

### *5.2.3 Generation Scheduling*

The generation sources, especially RES, are arranged in a good manner to follow and meet the daily load demand. These arrangements and planning depend on the daily load forecasting, the prediction of RES output power and the arrangement of the generation sources that will supply the load demand at critical conditions. Figure 5.5 presents the flowchart that explains the arrangement of generation sources to supply the demand.

The energy generated by CSP, PV, and conventional generators must satisfy Eq. (5.1) [6].

$$P_L = P_{CSP} + P_{PV} + P_{g,conv} \tag{5.1}$$

where,

$P_L$: Load power.
$P_{PV}$: PV system output power.

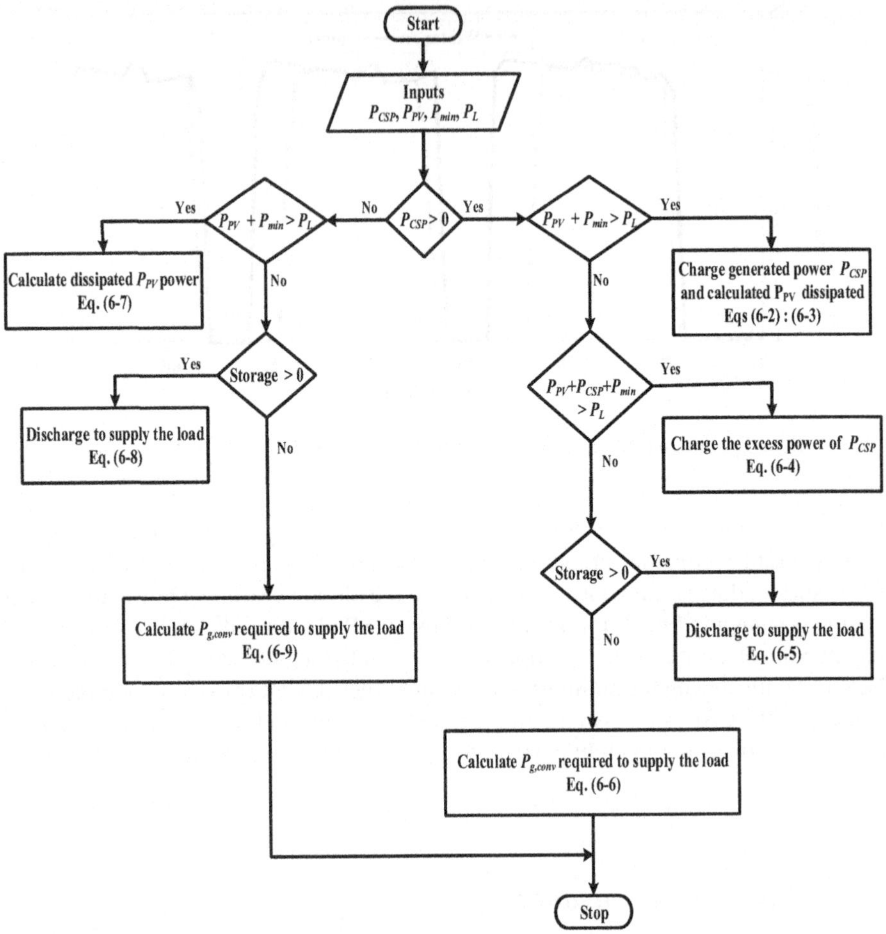


**Fig. 5.5** Generation sources planning to meet load demand

$P_{CSP}$: CSP system Output.
$P_{g,conv}$: Conventional generators output power.

For this hybrid system, there are two arrangement modes which are listed below [10, 11]:

**Mode 1:** the power generated by CSP > 0.

- If the electrical power generated by PV plus the minimum power of the conventional generators is higher than the load demand, the power generated by CSP will be charged and the dissipated PV power will be calculated as follows:

$$P_{ch} = P_{CSP} \tag{5.2}$$

$$P_{PV,dissp} = P_{PV} + P_{min,conv} - P_L \tag{5.3}$$

where,

$P_{ch}$: Charge power of the storage system.
$P_{min,conv}$: Minimum generation constraint.
$P_{PV,dissip}$: Curtailed PV power when the generation exceeds the demand and plants unable to reduce their output due to generators constraint.

- If the electrical power generated by PV, CSP and the minimum power of the conventional generators is higher than the load demand, the excess power will be charged.

$$P_{ch} = P_{CSP} + P_{PV} + P_{min,conv} - P_L \tag{5.4}$$

- If the electrical power generated by PV, CSP and the minimum power of the conventional generators is less than the load demand, check the thermal storage.

  When storage > 0

$$P_{disch} = -[P_L - (P_{CSP} + P_{PV} + P_{min,conv})] \tag{5.5}$$

where,

$P_{disch}$: discharge power of the storage system.

When storage < 0

$$P_{g,conv} = P_L - (P_{CSP} + P_{PV} + P_{min,conv}) \tag{5.6}$$

**Mode 2:** the power generated by CSP < 0.

- If the electrical power generated by PV plus the minimum power of the conventional generators is higher than the load demand, the dissipated PV power will be calculated.

$$P_{PV,dissp} = P_{PV} + P_{min,conv} - P_L \tag{5.7}$$

– If the electrical power generated by PV and the minimum power of the conventional generators is not sufficient to supply the load demand, check the thermal storage.

  When storage > 0

$$P_{disch} = -[P_L - (P_{PV} + P_{min,conv})] \tag{5.8}$$

  When storage < 0

$$P_{g,conv} = P_L - (P_{PV} + P_{min,conv}) \tag{5.9}$$

## 5.3 Simulation Results

The hourly generation of the two types of solar energy and the conventional generators is given for 90 h (about three days) as an example for analyzing the performance of CSP on the system. Also in this analysis, the minimum generation constraint of the conventional generators is 40 MW. For the first case study, Fig. 5.6a, b shows the output power and generation sharing of each source, respectively. It is observed that there is no problem at low penetration of PV system ($P_{PV} = 10\%$ of the total power).

While at a penetration level of $P_{PV} = 17\%$ of the total power as seen in Fig. 5.7a and in the case of the system does not contain electrical batteries storage, it is observed that the generation is more than the demand particularly at the middle of the day as shown in Fig. 5.7b. The increase in the power generation (dissipated PV power) causes the power quality problems. Therefore, this power should be reduced to meet the demand. This means there is some of the free solar power will be curtailed and lost. Consequently, the system profit will be reduced.

Therefore, in order to utilize the available solar power, the minimum generation constraint should be reduced to 35 MW as shown in Fig. 5.8. This solution causes another challenge for the conventional generators. Where the generators must have more flexibility to increase or decrease its output power to accommodate the power changes due to the load demand nature and large variability of the PV power, especially at high penetration level. The generated power increases sharply from 37 to 73 MW as shown in Fig. 5.8.

For the second case study, Fig. 5.9a shows the performance of two types of hybrid solar energy (PV and CSP). At a penetration level of solar energy about 23% of the total demand ($P_{PV} = 10\%$ and $P_{CSP} = 13\%$), it is observed that all the available PV power is utilized without the need to reduce the minimum generation constraint for the conventional generators. This result due to the ability of CSP to store the surplus power and dispatch it depending on the required demand as shown in Fig. 5.9a. Figure 5.10a presents another impact of CSP on PV system considering high penetration level about 30% of the demand ($P_{PV} = 17\%$ and $P_{CSP} = 13\%$) with minimizing

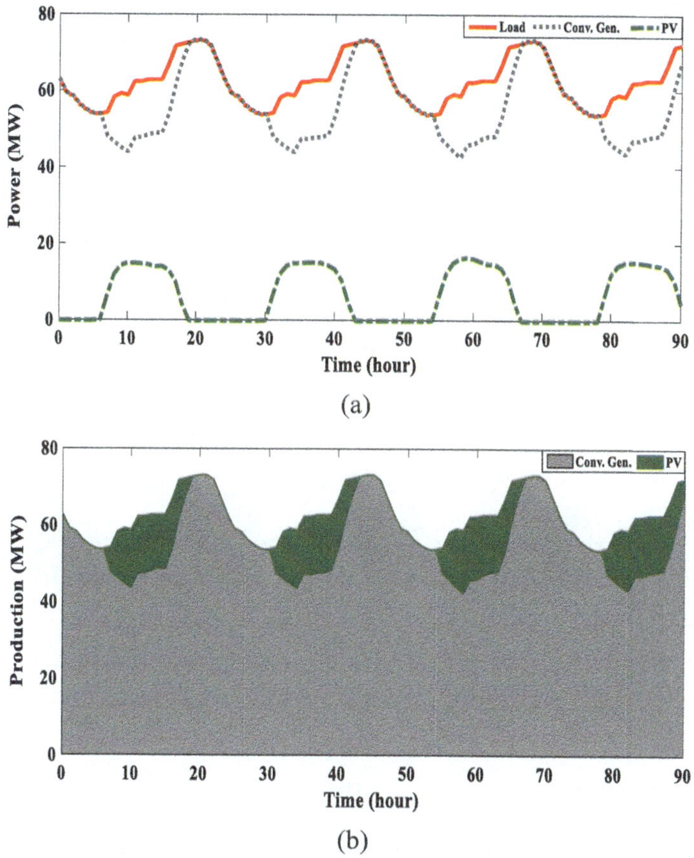


**Fig. 5.6** PV system at 10% penetration: **a** output power of generation sources and **b** sharing of generation sources

the generation constraint to 35 MW. As seen in Fig. 5.10b, the generator output power is ramped up from about 35 to 57.5 MW. While in the case of PV only, the output power is increased from about 37 to 73 MW as seen in Fig. 5.8. This means that the hybrid PV and CSP provides the overall system flexibility and the conventional generators became more flexible to increase and decrease its output power in order to meet the demand. Therefore, CSP usage enables more penetration of PV system. Table 5.1 presents the effects of CSP on the system penetration and overall flexibility at different cases.

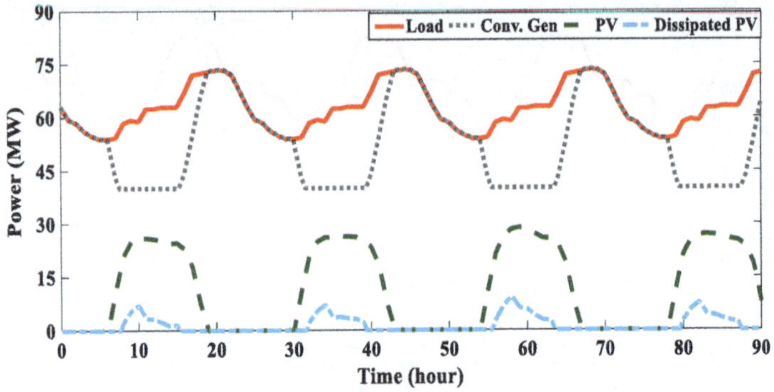


(a)

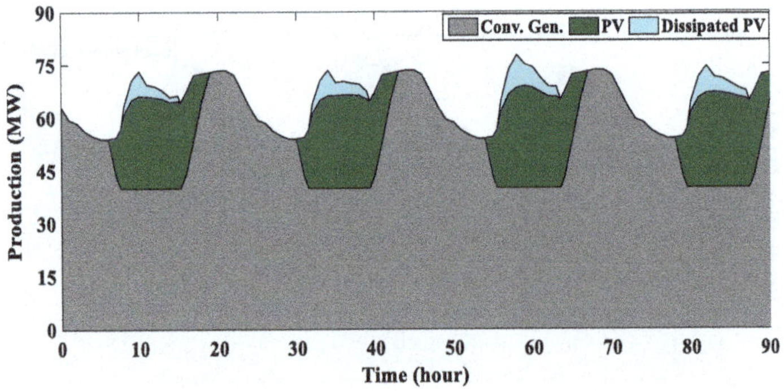


(b)

**Fig. 5.7** PV system at 17% penetration: (a) output power of generation sources and (b) sharing of generation sources

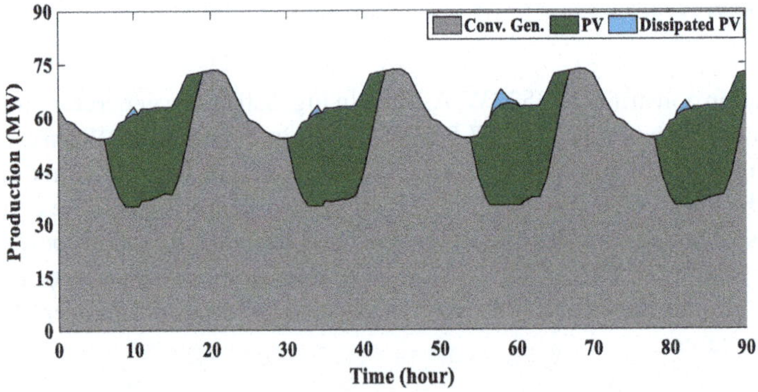


**Fig. 5.8** Generation sharing at 17% PV penetration and 35 MW minimum generation constraint

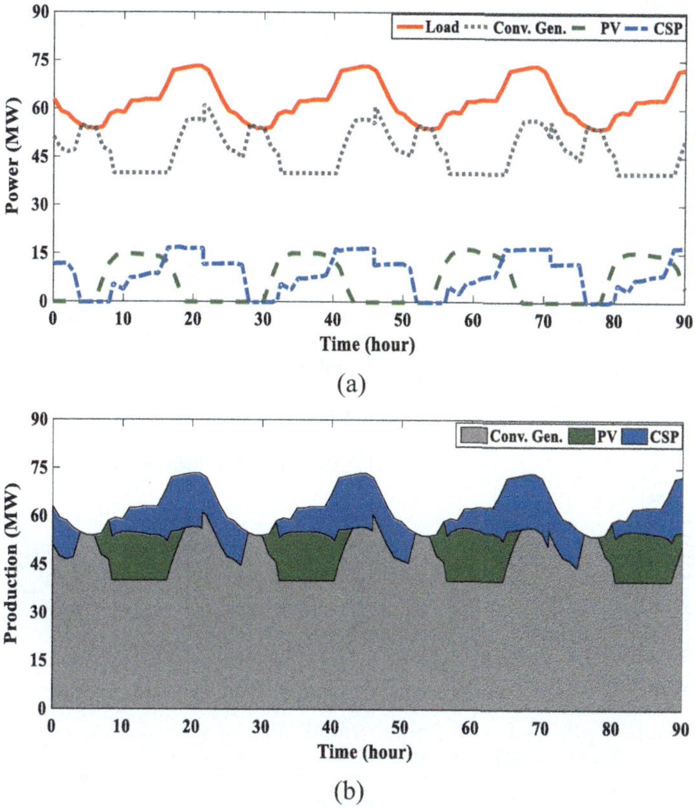


**Fig. 5.9** PV and CSP at 23% penetration: **a** output power of generation sources and **b** sharing of generation sources

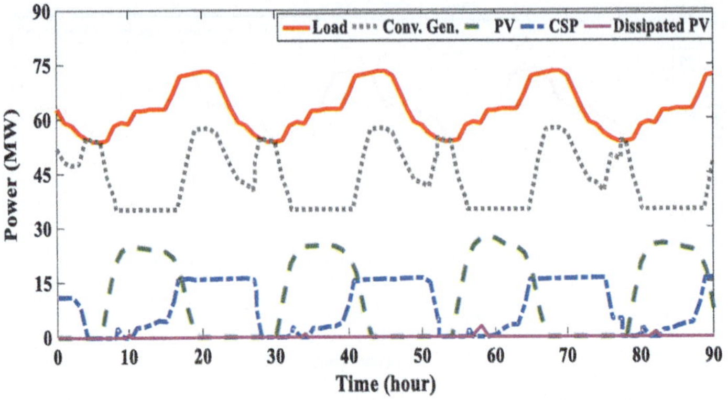


(a)

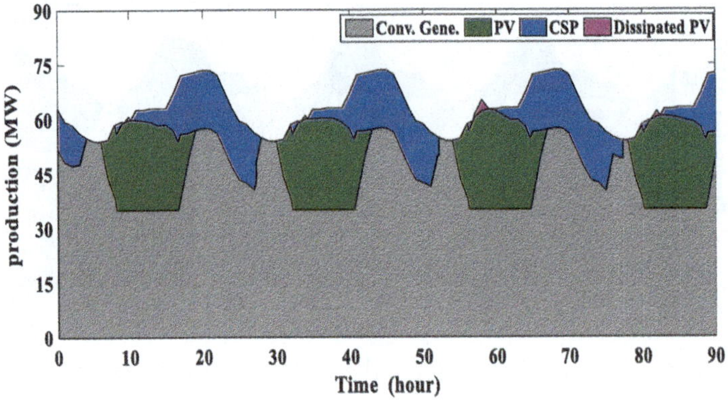


(b)

**Fig. 5.10** PV and CSP at 30% penetration and 35 MW minimum generation constraint: **a** output power of generation sources and **b** sharing of generation sources

**Table 5.1** System penetration and overall flexibility at different cases

| Case | $P_{PV}$ % | $P_{CSP}$ % | $P_{min}$ % | $P_{PV,Dissip}$ % | Ramp up of PG MW | System flexibility |
|---|---|---|---|---|---|---|
| 1 | 10 | - | 40 | - | 49–73 | - |
| 2 | 17 | - | 40 | 11 | 40–73 | - |
| 3 | 17 | - | 35 | 1.4 | 36–73 | Needs flexible generators |
| 4 | 10 | 13 | 40 | - | 40–57 | More flexible |
| 5 | 17 | 13 | 35 | 1.4 | 35–57 | Flexible |

# References


1. Eltawil MA, Zhao Z (2010) Grid-connected photovoltaic power systems: Technical and potential problems—a review. Renew Sustain Energy Rev 14(1):112–129
2. Mirhosseini M, Agelidis VG (2013) Interconnection of large-scale photovoltaic systems with the electrical grid: potential issues. In: IEEE international conference on industrial technology (ICIT), pp 728–733
3. Enslin JH, Alatrash H (2011) Distribution network impacts of high penetration of distributed photovoltaic systems. In: 21st international conference on electricity distribution, Frankfurt
4. Herrmann U, Kearney DW (2002) Survey of thermal energy storage for parabolic trough power plants. J SolEnergy Eng 124(2):145–152
5. Medrano M et al (2010) State of the art on high-temperature thermal energy storage for power generation. Part 2—Case studies. Renew Sustain Energy Rev 14(1):56–72
6. Elbaset AA, Hiyama T (2008) Optimal operation of electric hybrid WES/BS/DG system by neural network. In: International conference on electrical engineering, No 0–094
7. Fathy MS, Alaboudy AH, ELShater T (2012) Night operation of a photovoltaic system. In: 15th international middle east power systems conference, Alexandria, Egypt
8. NREL (2017) System Advisor Model (SAM): SAM Version 2017.1.17. Manual Release Date 2 June 2017
9. Blair NJ, Dobos AP, Gilman P (2013) Comparison of photovoltaic models in the system advisor model. National Renewable Energy Lab, NREL/CP-6A20–58057
10. Elbaset AA (2006) Study of interconnecting issues of photovoltaic/wind hybrid system with electric utility using artificial intelligence, PhD thesis. Minia University
11. Elbaset AA, Suryoatmojo H, Hiyama T (2010) Genetic algorithm based optimal sizing of PV-diesel-battery system considering CO2 emission and reliability. Int J Innov Comput Inf Control 6(10):4631–4649

# Chapter 6
# Economic Study of Solar Energy Systems

**Abstract** Global installed capacity of renewable energy technologies is growing rapidly. Hence, the technology assessment of energy production technologies is often computed as financial cost. Competition among alternative renewable technologies has increased substantially over the past few years, due to downward cost trends within each technology that have resulted from policy support and financial incentives. This chapter presents the results of the relationship between the energy price generated by the CTR plant with changing the number of storage hours (Ts), solar multiple, and also with the changing capacity of the station. Also, this chapter introduces program for optimal cost and LCOE of CTR system, PV and CTR/PV hybrid solar system. The computer program has been designed to determine optimum design parameters of PV and CTR for the system under study. The decision from the computer program is based on minimum price of the generated kWh from the system. Finally, the objective of this chapter is to research whether or not a solar PV system is more economical compared to the CTR system. The systems being considered in this study are in Aswan, Egypt as this region has hot and clear weather.

**Keywords** Cost analysis · System advisor model (SAM) · Electricity price · Levelized cost of electricity (LCOE) · Direct capital cost (DCC) · Indirect capital cost (ICC) · Operation and maintenance cost

## 6.1 Introduction

Global installed capacity of renewable energy technologies is growing rapidly. Hence, the technology assessment of energy production technologies is often computed as financial cost. The US Department of Energy and the National Renewable Energy Laboratory have been aggregating data on cost estimates for electricity generation in an online application. In this database, four main metrics exist to assess the cost of, especially, electricity generating infrastructure investment [1]: capital cost, operating costs and levelized cost of electricity (LCOE). As the still immature solar energy market has grown we have learned more about different technologies


I. Moukhtar et al., *Solar Energy*, Power Systems,
https://doi.org/10.1007/978-3-030-61307-5_6

and their ideal application. On the one hand, the flat panel PV cells, typically made of silicon, are the best form of solar technology, while the CSP industry is still at its infancy [2].

Competition among alternative renewable technologies has increased substantially over the past few years, due to downward cost trends within each technology that have resulted from policy support and financial incentives. However, comparisons of CSP with thermal storage with competing renewable technologies that focus only on differences in the LCOE are incomplete. This is because they do not capture the potentially significant differences in economic benefits when comparing renewable resources that have substantially different production characteristics [3, 4]. CSP with thermal energy storage is shown to be much more competitive when the comprehensive net costs of the CSP plant are compared to wind or PV. These net costs include the long-term energy, ancillary service and capacity benefits, and can be reasonably shown to provide an additional \$30–60/MWh, or even higher, of benefits when compared to a PV plant with equal annual energy production in high renewable penetration scenarios [4]. Hence, there are combined efforts between the two solar technologies that could result in higher total solar capacity value and less solar curtailment as PV penetration increases. There are also opportunities for CSP with thermal storage in remote locations to provide operational needs that cannot be cost-effectively provided by other renewable solutions. This chapter introduces a program for optimal cost and LCOE of CTR system, PV and CTR/PV hybrid solar system. The computer program has been designed to determine optimum design parameters of PV and CTR for the system under study. The decision from the computer program is based on minimum price of the generated kWh from the system.

The objective of this study is to research whether or not a solar PV system is more economical compared to the CTR system. The study analyzes this by performing a cost–benefit analysis for the lifetime of the solar systems. The systems being considered in this study are installed in Aswan, Egypt as this region has hot and clear weather.

## 6.2 Methodology of Cost Analysis of CTR/PV Systems

### *6.2.1 System Advisor Model*

System Advisor Model (SAM) is employed to estimate the performance and current/future costs for renewable energy such as PV and CSP electricity generation systems [5]. It incorporates modules that estimate the performance of different PV and CSP systems based on design parameters and climate files that include solar and weather data for the selected location. Also, SAM includes algorithms to estimate the LCOE based on a variety of selectable financial and incentive assumptions. Essential inputs of the LCOE calculations include the estimated installed cost and operating cost of the technology [6, 7]. SAM models grid-connected PV systems that consist of a PV array and inverter. The array can be made up of flat-plate or concentrating photovoltaic modules with one-axis, two-axis, or no tracking [8].

### 6.2.2 Calculation of Electricity Price

The levelized cost of electricity "LCOE" is an important metric which can be used to compare the economic competitiveness of different electricity generation systems [9]. The LCOE analysis can determine the benefits and drawbacks of various energy systems, which tell what is the most viable system to implement. It defines as the average cost per kWh of useful electrical energy produced by the plant. LCOE is calculated by a simplified methodology proposed by IEA as follows [10–12]:

$$LCOE = \frac{CRF \times K_{\text{invest}} + K_{O\&M} + K_{fuel}}{E_{e,g}} \tag{6.1}$$

$$K_{\text{invest}} = DCC + ICC \tag{6.2}$$

$$K_{O\&M} = F_C + V_C \tag{6.3}$$

where,

$LCOE$ Levelized cost of electricity.
$K_{\text{invest}}$ Total plant investment.
$K_{\text{fuel}}$ Annual fuel cost (CSP plants have virtually zero fuel costs, $K_{\text{fuel}} = 0$.
$K_{O\&M}$ Annual operation and maintenance.
$DCC$ Direct capital cost.
$ICC$ Indirect capital cost.
$E_{e,g}$ Annual energy production.
$F_C$ Fixed cost by capacity.
$V_C$ Variable cost by generation.

The capital recovery factor is given by [11, 13]:

$$CRF = \frac{k_d(1 + k_d)^m}{(1 + k_d)^m - 1} + k_{insur} \tag{6.4}$$

where,

CRF Capital recovery factor.
$k_d$ Discount rate.
$m$ Analysis period.
$k_{insur}$ Annual insurance rate.

#### 6.2.2.1 LCOE Calculation of CTR System

1. **Direct capital cost.**

It represents an expense for a specific piece of equipment or installation service that applies in year zero of the cash flow. It comprises the following costs [8, 10].

(a) Heliostat field.

A cost per square meter of total reflective area of the heliostat field to account for expenses related to installation of the heliostats, including heliostat parts, field wiring, drives, labor, and equipment.

$$HFC = A_{\text{mirror}} \times M_C \tag{6.5}$$

where,

*HFC* Heliostat field cost.
$M_C$ Mirror cost per unit area.

(b) Site improvements.

A cost per square meter of total reflective area of the heliostat field to account for expenses related to site preparation and other equipment not included in the heliostat field cost category.

$$SIC = A_{\text{mirror}} \times \text{SPC} \tag{6.6}$$

where,

*SIC* Site improvements cost.
SPC Site preparation cost per unit area.

(c) Total tower cost [8]:

$$TTC = FTC \times e^{TCSE\left(\frac{h_{t,SM} - h_{rec}}{2} + \frac{h_{hel}}{2}\right)} \tag{6.7}$$

where,

*TTC* Total tower cost.
*FTC* Fixed tower cost, which is a fixed cost to account for tower construction, materials and labor costs.
*TCSE* Tower cost scaling exponent, which defines the nonlinear relationship between tower cost and tower height.

$h_{rec}$ Receiver height.
$h_{hel}$ Heliostat height.

(d) Total receiver cost [8].

$$TRC = RRC\left(\frac{A_{rec}}{A_{ref,rec}}\right)^{RCSE} \quad (6.8)$$

where,

*TRC* Total receiver cost.
*RRC* Receiver reference cost. The cost per receiver reference area to account for receiver installation costs, including labor and equipment.
*RCSE* Receiver cost scaling exponent, which defines the nonlinear relationship between receiver cost and receiver area based on the reference cost conditions provided.
$A_{rec}$ Receiver area.
$A_{ref,\ rec}$ Receiver reference area; the receiver area on which the receiver reference cost is based.

(e) Thermal energy storage cost.

Cost per thermal megawatt-hour of storage capacity to account for the installation of a TES system, including equipment and labor.

$$TESC = P_{tes} \times C_{TS} \quad (6.9)$$

$$E_{tes} = P_{th,d} \times T_S \times \eta_{ts} \quad (6.10)$$

where,

*TESC* Thermal energy storage cost.
$C_{TS}$ Cost of thermal storage per $kWh$th.
$E_{tes}$ Designed stored thermal energy.
$T_S$ Storage hours.
$\eta_{ts}$ Storage system efficiency.

(f) Power cycle.

A cost per electric kilowatt of power cycle gross capacity related to installation of the PB components, including labor and equipment.

$$PCC = P_{e,d} \times PBC + P_{e,d} \times BPC \quad (6.11)$$

where,

*PCC* Power cycle cost.
*BPC* Balance of plant (\$/kWe), which is the cost per electric kilowatt of power cycle gross capacity related to installation of the balance-of-plant components and controls, and construction of buildings, including labor and equipment.
*PBC* Power block cost per kW.

(g) Contingency cost.

A percentage of the sum of the above costs to account for expected uncertainties in direct cost estimates.

$$CC = k_{con}(HFC + SIC + TTC + TRC + TESC + PCC) \tag{6.12}$$

where,

*CC* Contingency cost.
$k_{con}$ Constant value as a percent of subtotal cost.

Hence,

$$DCC = HFC + SIC + TTC + TRC + TESC + PCC + CC \tag{6.13}$$

where,

*DCC* Direct capital cost.

2. **Indirect capital cost.**

An indirect cost is typically one that cannot be identified with a specific piece of equipment or installation service [8, 10].

(a) Total land area cost.

$$TLAC = A_{land} \times L_C \tag{6.14}$$

where,

*TLAC* Total land area cost.
$L_C$ Land cost per unit area.

(b) Engineer-procure-construct and owner costs.

They are associated with the design and construction of the project such as permitting, royalty payments, consulting, management or legal fees, geotechnical and environmental surveys, interconnection costs, spare parts inventories, commissioning costs, and the owner's engineering and project development activities.

$$EPC\&OC = k_{EPC} \times DCC \tag{6.15}$$

where,

| | |
|---|---|
| $EPC\&OC$ | Engineer-procure-construct and owner costs. |
| $k_{EPC}$ | Constant value as a percent of the direct cost. |

(c) Sales tax.

$$ST = \text{sales tax rat} \times \text{percentage of direct cost} \times DCC \tag{6.16}$$

where,

$ST$ Sales tax.

Hence,

$$ICC = TLAC + EPC\&OC + ST \tag{6.17}$$

where,

$ICC$ Indirect capital cost.

3. **Operation and maintenance cost** [8, 10].

It represents annual expenditures on equipment and services that occur after the system is installed. These costs are specified as follows:

- Fixed cost by capacity ($/kW-year).
  A fixed annual cost proportional to the system's rated or nameplate capacity.
- Variable cost by generation ($/MWh).
  A variable annual cost proportional to the system's total annual electrical output in AC megawatt-hours.

#### 6.2.2.2 LCOE Calculation of PV System

The LCOE of PV system is determined by dividing the summation of the total yearly expenses of PV system by the expected yearly energy generated. The following equations are used to determine the LCOE [8].

$$TMC = N_{pv} \times \text{module cost} \times \text{module power} \tag{6.18}$$

where,

*TMC* Total modules cost.
$N_{pv}$ Total number of modules.

$$TIC = N_{inv} \times P_{inv} \times CoI \tag{6.19}$$

where,

*TIC* Total inverter cost.
$N_{inv}$ Total number of inverter.
$P_{inv}$ Rated power of inverter.
*CoI* Cost of Inverter.

$$BoS = BSC \times \text{Nameplate capacity} \tag{6.20}$$

where,

*BSC* Balance of system cost, $/$W_{dc}$.

$$IL = ILC \times \text{Nameplate capacity} \tag{6.21}$$

where,

*IL* Installation labor.
*ILC* Installation labor cost, $/$W_{dc}$.

$$\text{Contingency} = \text{percent of} \, (TMC + TIC + BOS + IL) \tag{6.22}$$

$$\text{Total direct cost of PV} = (TMC + TIC + BOS + IL + \text{Contingency}) \tag{6.23}$$

$$\text{ICC of PV system} = PES + EDO + GI + LCOPV + \text{PV total sale tax} \tag{6.24}$$

where,

*PES* Permitting—Environmental Studies.
*EDO* Engineering and developer overhead.

*GI* Grid interconnection.
*LCoPV* Land cost of PV system.

$$LCoPV = LAC + LPC \tag{6.25}$$

where,

*LAC* Land area cost of PV system, \$/acre.
*LPC* Land preparation cost of PV system, \$/$W_{dc}$.

$$\text{PV total sale tax} = STR \times PDC \times PVDC \tag{6.26}$$

where,

*STR* Sales tax rate of PV system.
*PDC* Percentage of Direct Cost of PV system.
*PVDC* Direct cost of PV system.

$$OMCPV = FCPV + VCPV \tag{6.27}$$

where,

*OMCPV* Operation and maintenance costs of PV system.
*FCPV* Fixed cost by capacity of PV system, \$/kW-yr.
*VCPV* Variable cost by generation of PV system, \$/MWh.

## 6.3 Case Study

This study shows and discusses the results of cost model for solar energy technologies, which explained in previous section. Therefore, the study examines the cost of two leading forms of solar power generation, PV and CTR technologies with different penetration levels and MW capacities as a representative case in Aswan site. It will attempt to determine whether PV or CTR is more effective and sustainable taking in consider prices of each power source.

## 6.4 Results and Discussion

This study was carried out in the area that has abundant solar energy available for a nearly whole year and hot and clear weather like Aswan city. This will be useful for energy planning and developing new strategies for solar power systems implementation. On the basis of the cost model described in chapter two, the calculation results of the LCOE for the two plants are presented in 2017 and its future evolution over the period of 2025 and 2030. Moreover, the CF and installation cost of the plant were calculated. All of this chapter's results depend on the input parameters that exist in chapter two.

### *6.4.1 CTR System Results*

In this section, the performance and economic analysis of a 10–100 MW CTR power plant is proposed. The objective of this study is to find the optimal size and number of storage hours that is more economical for the CTR system. The proposed program is employed to estimate the current costs and LCOE for CTR technology. The inputs listed in Table 6.1 were applied for calculating the costs and LCOE in this study at the year 2017. The cost and CF for this technology are predicted at 10 MW and 100 MW with different values of SM and $T_S$ as exhibited in Table 6.1, 6.2, 6.3, 6.4 and Table 6.5, respectively.

Figure 6.1 demonstrates the variation of CF for 10 MW and 100 MW capacity with various SM and $T_S$. It can be concluded that providing thermal energy without sufficient SM is not beneficial. For example at $T_S = 3$, it clear that the increasing SM beyond 2 is not suitable for economic consideration.

Figure 6.2 presents a combination of SM and thermal storage capacity for the two plant capacities. In the case of no storage, i.e., $T_S = 0$, the lowest LCOE occurs at SM

**Table 6.1** Cost input data used in all analyses in 2017 [14]

| Item | Value | Unit | Item | Value | Unit |
|---|---|---|---|---|---|
| $M_C$ | 145 | $/m$^2$ | $k_{con}$ | 7 | % |
| $SPC$ | 16 | $/m$^2$ | $L_C$ | $10 \times 10^3$ | $/acre |
| $FTC$ | $3 \times 10^6$ | $ | $k_{EPC}$ | 13 | % |
| $TCSE$ | 0.0113 | - | *Sales tax rate* | 5 | % |
| $RRC$ | $103 \times 10^6$ | $ | $K_{fuel}$ | 0 | $ |
| $RCSE$ | 0.7 | – | $F_C$ | 65 | $/kW-yr |
| $A_{ref,rec}$ | 1571 | m$^2$ | $V_C$ | 4 | $/MWh |
| $STC$ | 24 | $/kWh$_{th}$ | $k_d$ | 8 | % |
| $PBC$ | 1100 | $/kW$_e$ | $m$ | 25 | year |
| $BPC$ | 340 | $/kW$_e$ | $k_{insur}$ | 1 | % |

**Table 6.2** Variation of costs and LCOE with SM and TS for 10 MW capacity

| 10 MW | | | | | | | | |
|---|---|---|---|---|---|---|---|---|
| SM | Cost (Million $) | | | | LCOE (cents/kwh) | | | |
| | $T_S = 0$ | $T_S = 3$ | $T_S = 6$ | $T_S = 9$ | $T_S = 0$ | $T_S = 3$ | $T_S = 6$ | $T_S = 9$ |
| 1 | 47.82 | 51.2 | 53.26 | 54.51 | 30.77 | 34.76 | 36.69 | 37.66 |
| 1.5 | 55.06 | 57.98 | 59.65 | 61.5 | 23.07 | 21.59 | 22.44 | 23.36 |
| 2 | 62.62 | 65.41 | 67.61 | 69.66 | 25.24 | 17.98 | 17.56 | 18.1 |
| 2.5 | 70.37 | 73.07 | 75.07 | 77.03 | 27.28 | 18.83 | 15.59 | 15.26 |
| 3 | 79.02 | 81.54 | 84.03 | 85.54 | 29.13 | 20.1 | 16.38 | 14.21 |
| 3.5 | 85.90 | 88.88 | 91.16 | 93.09 | 31.05 | 21.35 | 17.28 | 14.64 |
| 4 | 94.33 | 97.06 | 98.71 | 100.7 | 33.33 | 22.91 | 18.38 | 15.51 |

**Table 6.3** Variation of CF with SM and $T_S$ for 10 MW capacity

| 10 MW | | | | |
|---|---|---|---|---|
| SM | CF (%) | | | |
| | $T_S = 0$ | $T_S = 3$ | $T_S = 6$ | $T_S = 9$ |
| 1 | 16.49 | 15.39 | 15.06 | 14.94 |
| 1.5 | 24.79 | 27.68 | 27.25 | 26.84 |
| 2 | 25.18 | 36.93 | 38.93 | 38.71 |
| 2.5 | 25.7 | 38.73 | 48.14 | 50.31 |
| 3 | 26.58 | 39.87 | 50.46 | 59.34 |
| 3.5 | 26.8 | 40.43 | 51.34 | 62.01 |
| 4 | 27.09 | 40.69 | 51.74 | 62.71 |

**Table 6.4** Variation of costs and LCOE with SM and $T_S$ for 100 MW capacity

| 100 MW | | | | | | | | |
|---|---|---|---|---|---|---|---|---|
| SM | Cost (Million $) | | | | *LCOE* (cents/kWh) | | | |
| | $T_S = 0$ | $T_S = 3$ | $T_S = 6$ | $T_S = 9$ | $T_S = 0$ | $T_S = 3$ | $T_S = 6$ | $T_S = 9$ |
| 1 | 341.63 | 363.92 | 385.77 | 407.96 | 19.03 | 20.21 | 21.2 | 22.37 |
| 1.5 | 411.67 | 433 | 454.85 | 477.1 | 16.54 | 14.71 | 15.37 | 15.96 |
| 2 | 482.54 | 504.33 | 522.64 | 544.85 | 18.38 | 13.66 | 12.67 | 13.11 |
| 2.5 | 550.86 | 572.15 | 593.91 | 613.93 | 19.92 | 14.64 | 12.19 | 11.5 |
| 3 | 622.26 | 644.32 | 664.62 | 685.86 | 21.84 | 15.95 | 13.09 | 11.31 |
| 3.5 | 692.87 | 715.91 | 734.85 | 754.86 | 23.95 | 17.38 | 14.17 | 12.11 |
| 4 | 761.55 | 781.92 | 804.96 | 827.39 | 26.14 | 18.71 | 15.25 | 13 |

Table 6.5 Variation of CF with SM and $T_S$ for 100 MW capacity

| 100 MW | | | | |
|---|---|---|---|---|
| SM | CF (%) | | | |
| | $T_s = 0$ | $T_s = 3$ | $T_s = 6$ | $T_s = 9$ |
| 1 | 20.5 | 20.28 | 20.11 | 19.97 |
| 1.5 | 27.4 | 32.16 | 31.97 | 31.96 |
| 2 | 27.9 | 39.26 | 43.7 | 43.64 |
| 2.5 | 28.66 | 40.57 | 50.58 | 55.26 |
| 3 | 28.88 | 41.06 | 51.68 | 61.63 |
| 3.5 | 28.82 | 41.15 | 51.94 | 62.53 |
| 4 | 28.61 | 41.18 | 52.11 | 62.95 |

of 1.5 for 10 MW and 100 MW. For a given storage capacity, as the SM increases, both the installation costs and CF (i.e., electricity output) increase. The interaction of these factors causes LCOE to decrease as SM increases from one, but at some point the increased cost overcomes the benefit of increased electric energy output, and the LCOE begins to increase with SM.

CTR technology can be deployed with large amounts TES and various SM; yielding high CF while maintaining an optimum LCOE. So, the optimal design of combination SM and $T_S$, at which LCOE is minimum, for the two plant capacities is described in Figs. 6.3, 6.4, 6.5 and 6.6. It can be seen that the increase in plant capacity has positive impact to decrease the cost of kWh. It appears from Fig. 6.7 that the energy price decreases notably when the plant capacity is increased until 40 MW. While, at large capacity, the LCOE decreases slowly when the plant capacity increases.

Table 6.6 shows the optimal design and cost of CTR plant for various rated capacities at 2017. It can be noticed that the CTR plant with high capacity has low price of kWh as shown in Table 6.6 and Fig. 6.7; the price of kWh for100 MW capacity is 11.31 cents/kWh. While Table 6.7 describes the current cost of CTR system for 2017 and target cost for 2025 and 2030. The continuous cost reductions in solar technologies established cost targets for 2025 and 2030 that would make solar one of the lowest-cost sources of new electricity [15, 16]. For CTR system, the new target corresponds to an LCOE of 9.88 cents/kWh and 5.80 cents/kWh in 2025 and 2030, respectively.

### 6.4.2 PV System Results

In this section, the performance and economic analysis of a 10–100 MW solar PV power plant is proposed. The objective of this study is to find the optimal parameters design of PV system and to research whether or not a solar PV system is more

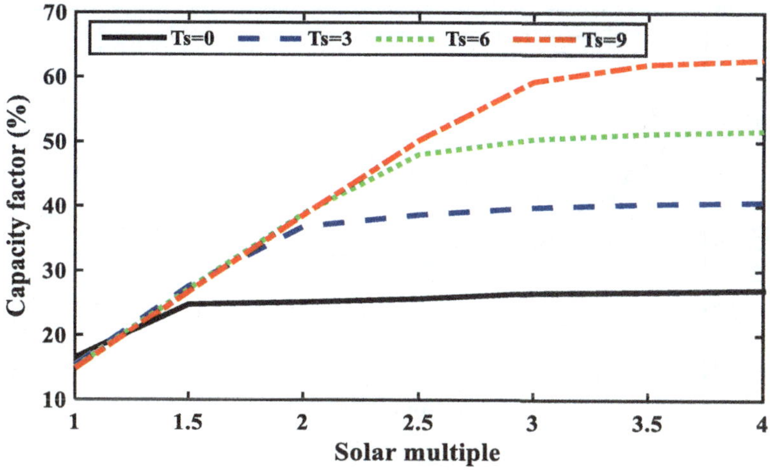


(a) 10MW plant capacity

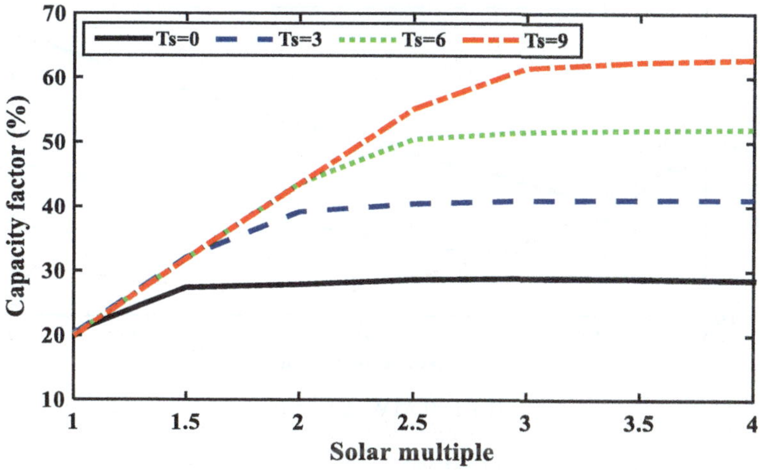


(b) 100 MW plant capacity

**Fig. 6.1** CF variation with various SM and $T_S$

economical compared to the CSP system. The inputs listed in Table 6.8 were applied for calculating the costs and LCOE of PV solar system in this study at the year 2017. While Table 6.9 shows the optimal design and cost of CTR plant for various rated capacities at 2017.

Figure 6.8 exhibits the variation of energy price with plant capacities. It can be found that the LCOE does not notable effect with large plant size like CTR power plant. Also, it can be notice that the solar PV plant with 100 MW represents the most economical one for Aswan site. The PV solar power plant with 100 MW can

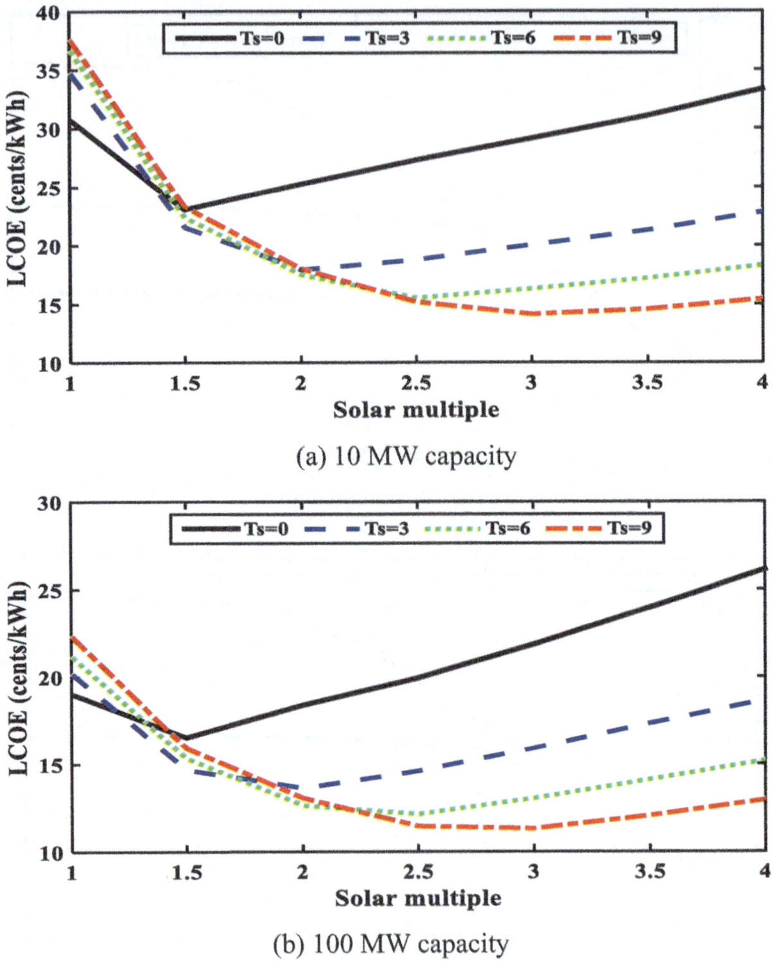


**Fig. 6.2** LCOE variation with various SM and $T_s$

generate electricity at 22% plant capacity factor which is less than the CF of CTR plant. This plant requires 322,416 PV modules, 100 inverters and 433 acres for total land area as described in Table 6.9. The price of kWh produced form 100 MW PV system is 7.15 cents/kWh. Table 6.10 demonstrates the current cost of PV system for 2017 and target cost for 2025 and 2030. As we mentioned before, the continuous cost reductions in solar technologies will make solar one of the lowest-cost sources of new electricity. Therefore, the new target of PV system corresponds to an LCOE of 5.27 cents/kWh and 4.66 cents/kWh in 2025 and 2030, respectively.

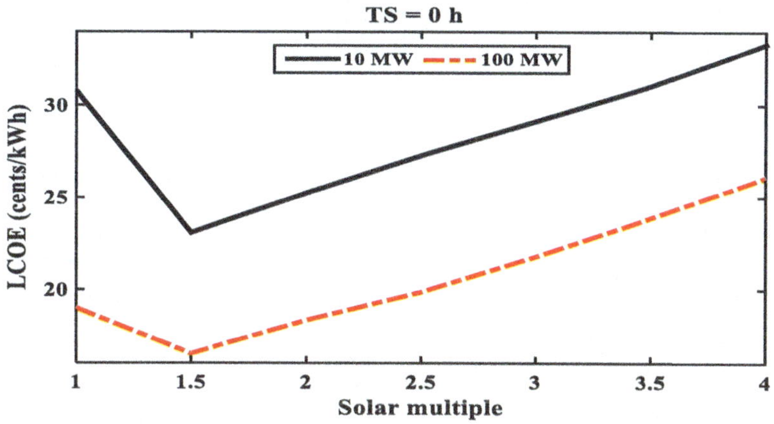


**Fig. 6.3** LCOE variation for different plant capacities with $T_S = 0$

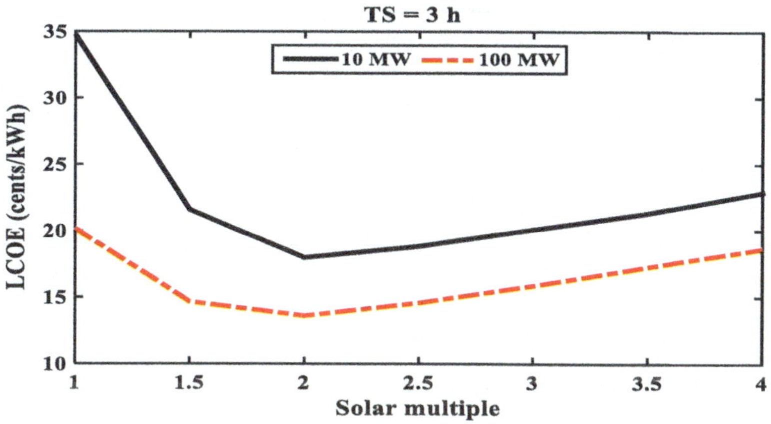


**Fig. 6.4** LCOE variation for different plant capacities with $T_S = 3$

### *6.4.3 CTR/PV Hybrid Solar System Results*

This section presents the results of the proposed program for optimal design of a CTR/PV hybrid system to be interconnected with electric grid. The proposed program has been designed to determine optimal design parameters of CTR/PV system. This hybrid system depends on dividing the load into two parts between CTR and PV. Design of CTR/PV hybrid system can be done by setting penetration ratio in the proposed program equal to 0,0.1,0.2,……0.9,1. The penetration level parameters determine the partition of the load. The results of optimal design parameters of CTR/PV system and LCOE has been summarized in Table 6.11.

From this table it can be seen that:-

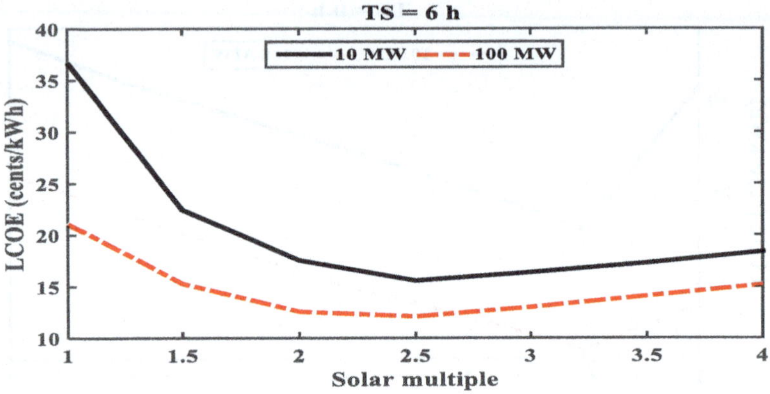


**Fig. 6.5** LCOE variation for different plant capacities with $T_S = 6$

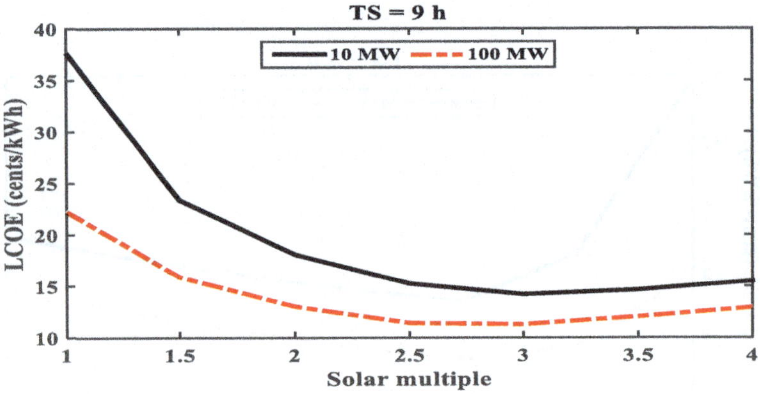


**Fig. 6.6** LCOE variation for different plant capacities with $T_S = 9$

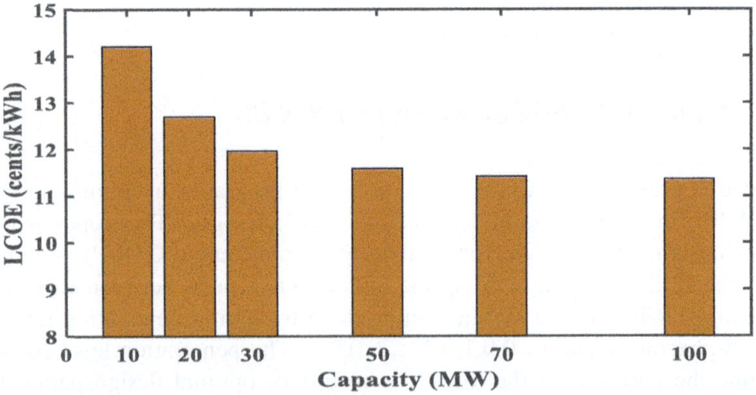


**Fig. 6.7** LCOE values for CTR plant at different capacities

**Table 6.6** Optimal design and cost of CTR plant with different MW for 2017

| MW | No of heliostats | Area (acres) | Installed cost (M$) | LCOE (cents/kW) | CF |
|---|---|---|---|---|---|
| 10 | 936 | 232 | 85.248 | 14.11 | 59.34 |
| 20 | 1805 | 372 | 154.917 | 13.05 | 60.8 |
| 30 | 2733 | 552 | 221.709 | 12.06 | 61.6 |
| 50 | 4581 | 878 | 355.104 | 11.63 | 61.8 |
| 70 | 6525 | 1304 | 484.798 | 11.38 | 61.7 |
| 100 | 9585 | 1996 | 685.866 | 11.31 | 61.63 |

**Table 6.7** Current and target LCOE of CTR plant

| Year | MW | Installed cost ($/W) | LCOE (cents/kWh) | CF |
|---|---|---|---|---|
| 2017 (current) | 100 | 6.8 | 11.31 | 61.6 |
| 2025 (goal) | 100 | 5.2 | 9.88 | 62.4 |
| 2030 (goal) | 100 | 3.3 | 5.80 | 63.1 |

**Table 6.8** Components cost of utility-scale PV system for 2017 [17]

| Module ($/$W_{dc}$) | Inverter ($/$W_{dc}$) | BoS ($/$W_{dc}$) | ILC ($/$W_{dc}$) | EDO ($/$W_{dc}$) | Conting (%) | OMCPV ($/kW-yr) | STR (%) |
|---|---|---|---|---|---|---|---|
| 0.35 | 0.06 | 0.25 | 0.14 | 0.1 | 0.03 | 20 | 5 |

**Table 6.9** Optimal design and cost of PV plant with different MW at 2017

| MW | No of modules | No of inverter | Area (acres) | Installed cost (M$) | LCOE (cents/kW) | CF |
|---|---|---|---|---|---|---|
| 10 | 32,232 | 10 | 43 | 12.602 | 8.16 | 22 |
| 30 | 96,720 | 30 | 130 | 35.655 | 7.79 | 22 |
| 50 | 161,208 | 50 | 217 | 55.777 | 7.49 | 22 |
| 70 | 225,696 | 70 | 303 | 75.325 | 7.33 | 22 |
| 100 | 322,416 | 100 | 433 | 103.353 | 7.15 | 22 |

- The lowest LCOE occurs at penetration level equals to 100% for PV and 0% for CTR which equals to 7.15 cents/kWh.
- The highest LCOE occurs at penetration level equals to 0 for PV and 100% for CTR which equals to 11.31 cents/kWh.
- The lowest CF occurs at penetration level equals to 100% for PV and 0% for CTR which equals to 22.
- The highest CF occurs at penetration level equals to 0% for PV and 100% for CTR which equals to 61.63.

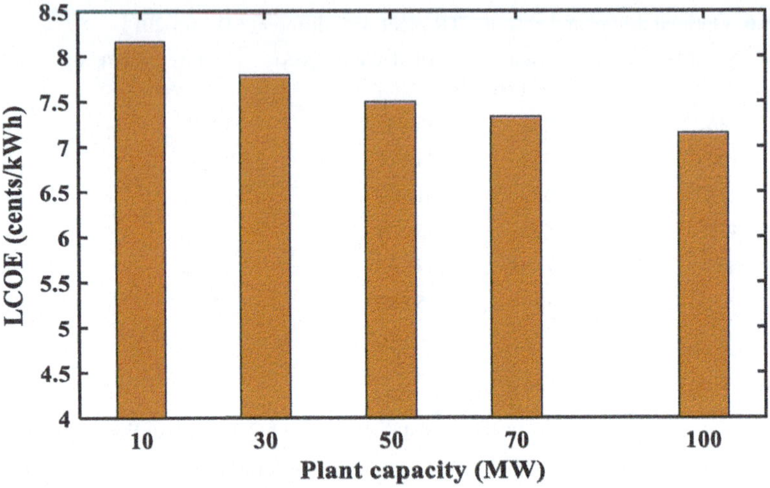


**Fig. 6.8** LCOE Values for PV plant at different capacities

**Table 6.10** Current and target LCOE of PV system

| Year | MW | Installed cost ($/W) | LCOE (cents/kWh) | CF |
|---|---|---|---|---|
| 2017 (current) | 100 | 1.11 | 7.15 | 22 |
| 2025 (goal) | 100 | 0.70 | 5.27 | 22 |
| 2030 (goal) | 100 | 0.51 | 4.66 | 22 |

**Table 6.11** Impact of penetration level on the optimum design of CTR/PV hybrid system

| Case | Penet. Level, PV | Penet Level, CTR | PV module No | Heliostat No | $CF_{hyb}$ | Installed $Cost_{hyp}$ (M$) | $LCOE_{hyp}$ (cents/kWh) |
|---|---|---|---|---|---|---|---|
| 1 | 0 | 1 | 0 | 9585 | 61.63 | 685.866 | 11.31 |
| 2 | 0.1 | 0.9 | 90,648 | 8512 | 52.35 | 648.376 | 10.89 |
| 3 | 0.2 | 0.8 | 181,308 | 7475 | 45.38 | 612.329 | 10.54 |
| 4 | 0.3 | 0.7 | 271,956 | 6525 | 40.07 | 571.976 | 10.12 |
| 5 | 0.4 | 0.6 | 362,616 | 5494 | 35.79 | 536.757 | 9.77 |
| 6 | 0.5 | 0.5 | 453,276 | 4581 | 32.44 | 500.406 | 9.39 |
| 7 | 0.6 | 0.4 | 543,924 | 3727 | 29.68 | 462.331 | 8.98 |
| 8 | 0.7 | 0.3 | 634,584 | 2733 | 27.26 | 425.131 | 8.62 |
| 9 | 0.8 | 0.2 | 725,244 | 1805 | 25.05 | 387,401 | 8.32 |
| 10 | 0.9 | 0.1 | 815,892 | 936 | 23.43 | 346.790 | 7.84 |
| 11 | 1 | 0 | 906,552 | 0 | 22 | 290.604 | 7.15 |

- If the hybrid system consists of CTR and PV, i.e. the LCOE value and CF increase with the increase of the penetration level of CTR system and vice versa.

## 6.5 Summary Results of Solar Energy Systems Cost

The results obtained from the financial assessment of the CTR and PV plants are shown in Fig. 6.9, where the LCOE presents values in 2017,which are 11.31 and 7.15 cents/kWh for CTR and PV, respectively. While the CF for CTR and PV are 61.6 and 22 as shown in Fig. 6.10, respectively. It can be seen that the price of kWh

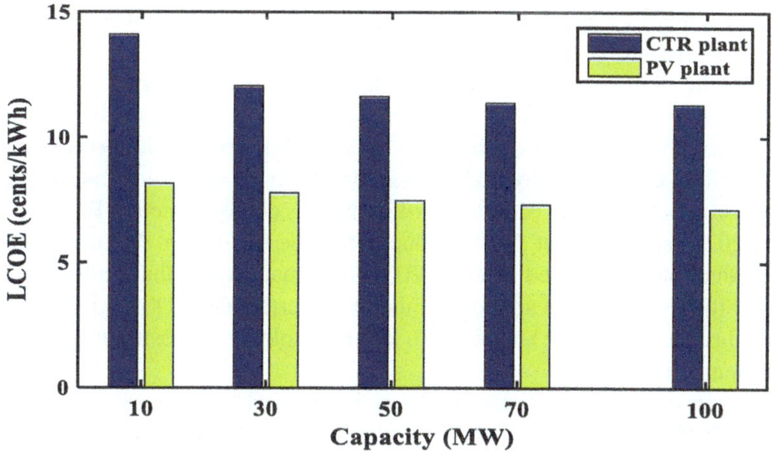


**Fig. 6.9** Comparison between LCOE of CTR and PV plants in 2017

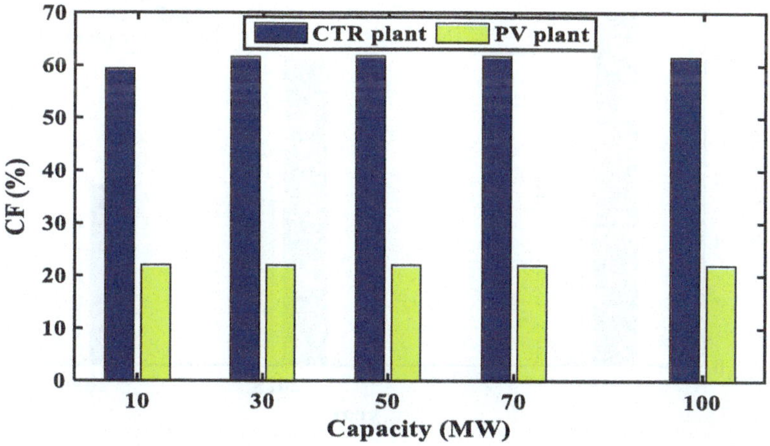


**Fig. 6.10** Comparison between CF of CTR and PV plants in 2017

for CTR system is much higher than that of PV system, especially at small plant size, which is mainly due to the following reasons:

- The industry of CTR is still at its infancy and therefore the plant components are very expensive compared to PV plant components.
- The used land area is very large compare with that used for PV plant as presented in Tables 6.6 and 6.9.
- CTR plants with TES tend to be significantly more expensive due to the storage system and the larger solar field.

Although the high price of energy production from CTR technology, it can be observed that its CF is still higher than PV plant owing to the use of TES system as illustrated in Fig. 6.10. So, it can shift the generation when the sun does not shine and/or the ability to maximize generation at peak demand times, which has positive effect on the electric grid reliability than PV system.

Finally, the analysis in the above study highlights that although CTR technology is commercially mature, it is far from mature from a cost perspective. In general, there is an opportunity for reducing the cost of electricity of solar power technologies, especially CTR plant, by 2025 and 2030 due to the technological innovations, increased competition, and economies of scales. As demonstrated in Fig. 6.11, from 2017 to 2025 there is about 26.3% reduction in the LCOE for PV plant and about 34.8% reduction is expected through 2030. The majority of that reduction can be attributed to the total hardware costs (module, inverter, and BOS). Also, there is an additional reduction can be attributed to labor and other soft costs such as sales tax, overhead, and net profit.

While approximately 12.6 and 48.7% reduction in LCOE was expected through 2025 and 2030 for CTR plant, respectively. It can be found that, the energy price drop in 2030 for CTR technology is significantly greater than PV due to technology

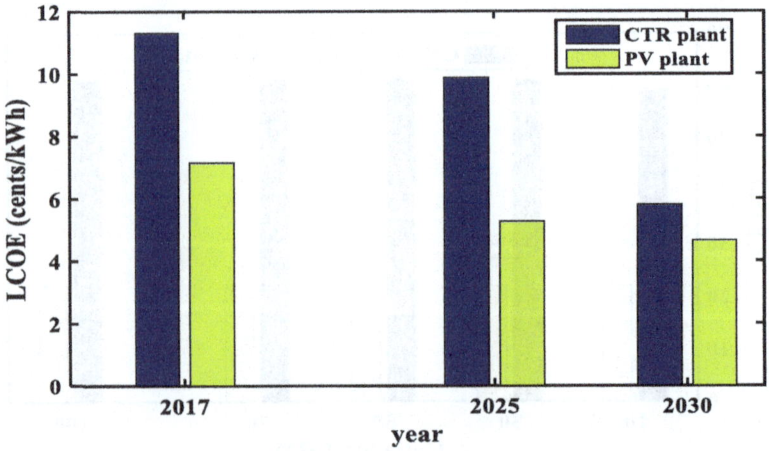


**Fig. 6.11** Comparison of current and target LCOE for PV and CTR plan

improvements, economies of scale, and other factors such as reduction of TES cost due to high operating temperature. Therefore, the CTR systems might become the technology of choice in the near future, because they can achieve very high temperatures with manageable losses by using molten salt as a heat transfer fluid. This will allow higher operating temperatures and steam cycle efficiency, and reduce the cost of TES by allowing a higher temperature differential. Their main advantage compared to PV systems is therefore that they could economically meet peak air conditioning demand and intermediate loads (in the evening when the sun isn't shining) in hot areas in the near future.

## References

1. Dale M (2013) A comparative analysis of energy costs of photovoltaic, solar thermal, and wind electricity generation technologies. Appl Sci 3(2):325–337
2. Banoni VA, Arnone A, Fondeur M, Hodge A, Offner JP, Phillips JKJCCJ (2012) The place of solar power: an economic analysis of concentrated and distributed solar power. Chem Cent J 6(1):S61–S11
3. Alliance CJ (2014) The economic and reliability benefits of CSP with thermal energy storage: literature review and research needs. CSP Alliance Report
4. Joskow P (2011) Comparing the costs of intermittent and dispatchable electricity generating technologies. Am Econ Rev 101(3):238–241
5. https://sam.nrel.gov/download.html
6. Turchi CS, Heath GA (2013) Molten salt power tower cost model for the system advisor model (SAM). National Renewable Energy Lab, Technical Report NREL/TP-5500–57625
7. Turchi C, Mehos M, Ho CK, Kolb GJ (2010) Current and future costs for parabolic trough and power tower systems in the US market. National Renewable Energy Lab, NREL/CP-5500–49303
8. NREL System Advisor Model (SAM): SAM Version 2017.1.17. Manual Release Date 2/6/2017
9. Musi R et al (2017) Techno-economic analysis of concentrated solar power plants in terms of levelized cost of electricity. AIP Conf Proc 1850(1):160018
10. Ramaswamy M et al (2012) Engineering economic policy assessment of concentrated solar thermal power technologies for India. Center for Study of Science Technology and Policy, Report CSTEP/E/7
11. Paal RP, Dersch J, Milow B (2003) European concentrated solar thermal road-mapping: roadmap document. ECOSTAR, SES6-CT-502578
12. Craig O, Brent A, Dinter F (2017) The current and future energy economics of concentrating solar power (CSP) in South Africa. S Afr J Ind Eng 28(3):1–14
13. Short W, Packey D, Holt T (1995) A manual for the economic evaluation of energy efficiency and renewable energy technologies. National Renawable Energy Laboratory publication, NREL/TP-462–5173
14. Turchi C (2017) Concentrating solar power: current cost and future directions. Available at https://www.wesrch.com/energy/pdfTR1L02000NXPF
15. Murphy C et al (2019) The potential role of concentrating solar power within the context of DOE's 2030 solar cost targets. National Renewable Energy Lab, Golden, Technical Report NREL/TP-6A20–71912
16. Taylor M, Ralon P, Ilas A (2016) The power to change: solar and wind cost reduction potential to 2025. International Renewable Energy Agency
17. Fu R et al (2017) US solar photovoltaic system cost benchmark: Q1 2017. National Renewable Energy Laboratory, Technical Report NREL/TP-6A20–68925

# Chapter 7
# Conclusions and Future Works

**Abstract** This chapter reports the main conclusions that can be drawn from the book and open the window for different research points in future work.

## 7.1 Conclusions

This book presents a proposed design of solar energy systems located in Aswan city in Egypt. As well as it introduces a proposed model of central tower receiver (CTR) power plant using ANN from a reasonably simplified model perspective. The simulation results using MATLAB/Simulink of the proposed CTR-ANN model were compared with those simulated by System Advisor Model (SAM) software. In addition, this work studies the effect of concentrated solar power (CSP) plant on the system reliability and penetration level of photovoltaic (PV) systems. Finally, this work analyzes the installed cost and price of energy production of solar energy systems. This work has achieved the following conclusions:

- A computer program has been proposed for optimal design of CTR, PV, and CTR/PV hybrid systems to be interconnected with electric grid. This computer program can be applied in any site of the world.
- For CTR system, the energy price (LCOE) decreases notably when the plant capacity is increased until 40 MW. While, at large capacity (>40 MW), the LCOE decreases slowly when the plant capacity increases.
- The prices of kWh produced form CTR system are 11.31, 9.88, and 5.8 cents/kWh for years 2017, 2025, and 2030, respectively.
- The prices of kWh produced form PV system are 7.15, 5.27, and 4.66 cents/kWh for years 2017, 2025, and 2030, respectively.
- Although the kWh price of CTR system is much higher than of PV system, its CF (about 61.6) is still higher than PV plant (about 22) owing to the use of TES system. So, it can shift the generation when the sun does not shine and/or the power is needed at peak demand times.
- The reduction in LCOE through 2030 for CTR plant is significantly greater than PV due to its technology improvements and other factors such as reduction of



I. Moukhtar et al., *Solar Energy*, Power Systems,
https://doi.org/10.1007/978-3-030-61307-5_7

TES cost. Therefore, the CTR systems might become the technology of choice in the near future.
- MLP neural network based on LM as a training algorithm with 40 neurons in the first hidden layer is the optimal topology compared to other algorithms. Hence, it can be used as an effective technique to precisely control the mass flow rate and subsequently the receiver outlet temperature regardless of the variations in the direct solar radiation and receiver inlet temperature.
- The simplified CTR-ANN model can be generalized as an adequate tool to predict and analyze the performance of CTR power plant in a simple and fixable manner. In addition, it can help in the design steps during project implementation.
- The simplicity and minimum required input data of our proposed model make it appropriate for evaluating the power system reliability by using MC method.
- Indeed, the use of CSP plant with TES reduce the high penetration problems of PV system and increases the overall system reliability.
- The use of CSP reduces the minimum generation constraint of the conventional generators that allows more penetration of the PV system.

## 7.2 Recommendation for Future Works

This book has opened the window for different research points. Some of them are listed here:

- For more accurate model, CTR plant losses will be taken into consideration in the future work.
- Use advanced optimization programs to determine the optimal distribution for heliostats locations.
- Despite the ANN model has been extensively applied to optimize, predict, and analyze the characteristics of different solar energy technologies, the analysis of CTR/TES performance using ANN model still requires more efforts.
- The cost factor and LCOE should be taken into consideration as a very important measure when study CSP effects on PV penetration levels in order to determine the optimal size of the hybrid CSP and PV systems.
- The price of energy produced from conventional generators should be taken into consideration when calculate LCOE value to get the accurate optimal size of the hybrid CSP and PV systems based on minimum LCOE.

# Appendix A
# Parameters of DI Method

Liu and Jordan showed that $r_d$ is well expressed by [1, 2].

$$r_d = \frac{\pi}{24}\left(\frac{\cos h_s - \cos h_{ss}}{\sin h_{ss} - \frac{\pi}{180} h_{ss} \cos h_{ss}}\right) \tag{A.1}$$

and

$$r_t = r_d\left(\frac{1 + q\left(\frac{a_2}{a_1}\right) A(h_{ss}) r_d \left(\frac{24}{\pi}\right)}{1 + \frac{q\left(\frac{a_2}{a_1}\right) B(h_{ss})}{A(h_{ss})}}\right) \tag{A.2}$$

The hour angles of sunset, $h_{ss}$, is given by:

$$h_{ss} = \cos^{-1}(\tan \mathrm{L}_{\mathrm{lat}} \cdot \tan\delta_s) \tag{A.3}$$

The factors $A(h_{ss})$, $B(h_{ss})$ and $q$ are calculated by the following equations:

$$A(h_{ss}) = \sin h_{ss} - h_{ss} \cos h_{ss} \tag{A.4}$$

$$B(h_{ss}) = h_{ss}\left(0.5 + \cos^2 h_{ss}\right) - 0.75 \sin(2h_{ss}) \tag{A.5}$$

$$q = \cos L - \cos\delta_s \tag{A.6}$$

$(a_2/a_1)$ is the atmospheric extinction effect is given as follow:

$$a_1 = 0.4134K_t + 0.61197K_t^2 - 0.01886K_t S_d + 0.00759S_d \tag{A.7}$$

$$a_2 = \mathrm{Max}\left(0.054, 0.28116 + 2.2475K_t - 1.7611K_t^2 - 1.84535 \sin h_o + 1.681 \sin^3 h_o\right) \tag{A.8}$$


I. Moukhtar et al., *Solar Energy*, Power Systems,
https://doi.org/10.1007/978-3-030-61307-5

The day length, $S_d$, (in hours) is obtained as:

$$S_d = \frac{24}{\pi} h_{ss} \tag{A.9}$$

The daily average clearness index, $K_t$, is given by:

$$K_t = \frac{\overline{H}_h}{H_o} \tag{A.10}$$

$H_o$ is the daily-average extraterrestrial irradiation on a horizontal surface and it is calculated as a function of the solar constant as:

$$H_o = \frac{24}{\pi} h_{ss} R_{cf} I_o \sin h_o \tag{A.11}$$

where $I_o$ is the solar constant; $I_o = 1367\ \mathrm{W/m^2}$
The monthly average sun–earth correction factor, $R_{cf}$, is given by:

$$\begin{aligned} R_{cf} = & 1.00011 + 0.034221\cos(B) + 0.00128\sin(B) \\ & + 0.000719\cos(2B) + 0.0000\sin(2B) \end{aligned} \tag{A.12}$$

The daily average solar elevation outside of the atmosphere is $h_o$, defined by

$$h_o = \sin^{-1}\left(\frac{qA(h_{ss})}{h_{ss}}\right) \tag{A.13}$$

## References

1. Kalogirou SA (2013) Solar energy engineering: processes and systems, 2nd edn. Academic Press
2. Goswami DY (2015) Principals of solar engineering, 3rd edn. CRC Press, Taylor & Francis Group

# Appendix B
# Heat Exchanger Calculation

The maximum heat transfer rate between the two fluids.

$$Q_{\max} = C_{\min}\left(T_{HTF,in} - T_{\text{steam},in}\right) \tag{B.1}$$

When the saturated water changed into saturated vapor, the heat capacitance is infinite. Therefore, the minimum capacitance equals the capacitance of the fluid of hot side as in the following equation.

$$C_{\min} = m_{HTF}C_{p,HTF} \tag{B.2}$$

At the reference full-load condition, the heat transfer effectiveness $\varepsilon$ and heat transfer unit of a shell and tube heat exchanger, which is the common type of heat exchanger used in the solar power plants, are calculated as the following [3, 4]:

$$\varepsilon_{ref} = 2\left\{1 + C_r + \sqrt{1 + C_r^2}\left(\frac{1 + e^{-NTU\sqrt{1+C_r^2}}}{1 - e^{-NTU\sqrt{1+C_r^2}}}\right)\right\}^{-1} \tag{B.3}$$

$$NTU_{ref} = \frac{-1}{\sqrt{1 + C_r^2}} ln\left(\frac{\frac{2}{\varepsilon} - 1 - C_r - \sqrt{1 + C_r^2}}{\frac{2}{\varepsilon} - 1 - C_r + \sqrt{1 + C_r^2}}\right) \tag{B.4}$$

Where capacitance ratio equals zero due to the maximum heat capacitance is infinite [4]:

$$C_r = \frac{C_{\min}}{C_{\max}} \approx 0 \tag{B.5}$$


I. Moukhtar et al., *Solar Energy*, Power Systems,
https://doi.org/10.1007/978-3-030-61307-5

Thus, ε and NTU are simplified as the following equations:

$$\varepsilon_{ref} = 1 - e^{NTU_{ref}} \tag{B.6}$$

$$NTU_{ref} = -\ln(1 - \varepsilon) \tag{B.7}$$

## References

1. Zohuri B (2017) Compact heat exchangers: selection, application, design and evaluation. Springer
2. Lee HS (2010) Thermal design: heat sinks, thermoelectrics, heat pipes, compact heat exchangers, and solar cells. Wiley

The manufacturer's authorised representative in the EU is Springer Nature Customer Service Centre GmbH, Europaplatz 3, 69115 Heidelberg, Germany. If you have any concerns regarding our products, please contact ProductSafety@springernature.com

Printed and bound by CPI Group (UK) Ltd, Croydon, CR0 4YY

30/09/2026

02225665-0008